T0304389

LUCID DYING

The New Science Revolutionizing How We Understand Life and Death

LUCID DYING

SAM PARNIA, MD, PHD

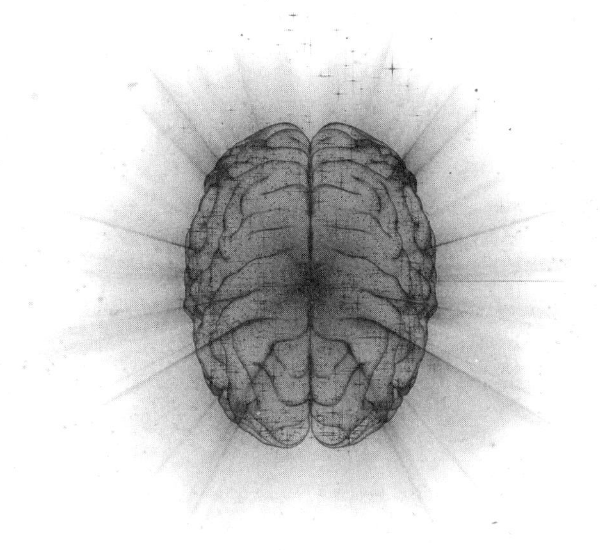

New York

Hachette Books
Hachette Book Group
1290 Avenue of the Americas
New York, NY 10104
HachetteBooks.com
Twitter.com/HachetteBooks
Instagram.com/HachetteBooks

First Edition: August 2024

Published by Hachette Books, an imprint of Hachette Book Group, Inc. The Hachette Books name and logo is a trademark of the Hachette Book Group.

The Hachette Speakers Bureau provides a wide range of authors for speaking events. To find out more, go to hachettespeakersbureau.com or email HachetteSpeakers@hbgusa.com.

Books by Hachette Books may be purchased in bulk for business, educational, or promotional use. For information, please contact your local bookseller or Hachette Book Group Special Markets Department at: special.markets@hbgusa.com.

Print book interior design by Amy Quinn

Library of Congress Cataloging-in-Publication Data has been applied for.

ISBNs: 9780306831287 (hardcover), 9780306831300 (ebook)

Printed in Canada

MRQ

Printing 1, 2024

CONTENTS

LUCID DYING

INTRODUCTION

WHETHER WE LIKE IT OR NOT, WE WILL ALL DIE, AND THE INEVitability of our death is one of only two certainties in life. As Benjamin Franklin famously wrote in 1789: "In this world nothing can be said to be certain, except death and taxes."* We know that we will die, yet aside from a minority of people for whom this reality becomes an overwhelming preoccupation, this thought takes a backseat to the daily preoccupations of life.

Nonetheless, the reality of our inevitable mortality and the importance of discovering what happens when we die become real when we lose a loved one, experience a personal tragedy—say, an accident or unexpected illness—or simply begin to age and come face-to-face with our own impending death. In those moments of confusion, fear, or acceptance, we begin to really think about questions that have plagued humanity for thousands of years: What is life and what is death? What happens when we die? Where does our consciousness—*our self*—go after death? Is there anything beyond?

Our beliefs about death and the importance we give to answering these questions are largely shaped by societal, personal, historical, religious, and cultural ideologies.

As a result, today most people's beliefs, including those of most scientists, are largely rooted in traditional and philosophical notions of death,

* Franklin is not the first to have used this idiom. In fact, many think it may have originated from Christopher Bullock's 1716 comedy play *The Cobler of Preston*, in which he states: "'Tis impossible to be sure of anything but Death and Taxes."

inherited from our ancestors over thousands of years. But what if what we believe about death is fundamentally wrong and the paradigm we have been operating in no longer exists? Perhaps life and death can no longer be considered as easily separable binary events. Most importantly, what if death itself is not the end we have thought and assumed? What if who we are—*our very consciousness and selfhood*—is not annihilated with death? And what if science can prove it? What will that mean for us, for society, for religion, for philosophy, for how we live our lives?

Death concerns all of us, yet mainstream science has traditionally avoided trying to answer it. Some of the most heavily funded scientific research today, everything from multibillion-dollar trips to Mars, space tourism, and rockets to the International Space Station, has little impact on our lives. The one area that will certainly affect all of us—exploring death and what happens when we die—has unfortunately been outright ignored by mainstream science.

The reason might simply be that scientists think death is something unassailable and impossible to study with scientific rigor. After all, how do we study a phenomenon that ultimately results in the loss of our subjects? If we view death as "the end"—rather than as a medical condition—it becomes easier to understand why people have assumed there is little point in furthering discussion or exploration.

Many scientists consider what happens after death as mere philosophical or theological speculation. I don't agree with this position and thankfully, today this no longer needs to be the case. Using sophisticated brain technology, we can peer inside the brains of people as they approach and then transition beyond death by recording and analyzing their brain waves. Advances in mathematics, artificial intelligence (AI) technology, and computing power are allowing us to objectively analyze realities about life and death that we never could have until recently. And this research is giving us the first real-time ability to study what happens to the brain and consciousness in people as they cross from life beyond the boundary of death.

The scientific story of death and what happens when we die, while of utmost importance and interest to us all, has evolved rapidly over the past few decades. In the same way that scientific progress is challenging our views across many other spheres of society, so too it is influencing our understanding of what happens with death. Death, once the purview of cultural, religious, philosophical, biological, anthropological, and metaphysical perspectives, is not as simple or straightforward as many people have perceived it to be. In fact, scientific discoveries around death and the postmortem period have clarified that these traditional concepts largely derived from ancient philosophical and historical notions are either outdated and frankly wrong, or at best inadequate and inaccurate. Yet many scientists are still swayed by their personal views and historical notions of what death is, rather than an assessment of evolving scientific realities.

That is why after almost thirty years of being on the frontlines of research and clinical practice in this field, I am writing *Lucid Dying* with the goal to synthesize the most important scientific discoveries and make them available to professionals and the general public alike. By piecing the various pieces of this intriguing puzzle together, much like an orchestra conductor, my hope is to help you appreciate the compellingly beautiful and cohesively scientific, yet largely untold, story of what happens to us when we traverse from life into death and beyond.

My interest in cardiac arrest resuscitation, the study of the brain and consciousness, and understanding what happens to people as they transition between life and death started quite early in my career. In 1994, I was a visiting final-year medical student at Mount Sinai Hospital in New York. There, I met a very amiable elderly man with a loving family alongside him in the emergency room, with what seemed a non-life-threatening medical condition. Around half an hour later, the overhead emergency alarm system went off. Someone's heart had suddenly stopped—meaning there was a cardiac arrest—in the emergency department. I ran to find a sea of doctors in white coats frantically working. They were pounding up and down on the chest of what looked like a dead body: the corpse was covered in blood, and there was a frenzy of activity everywhere.

Those lifesaving resuscitation attempts went on for more than an hour. I watched as this person received round after round of emergency treatments to try to restart his heart. At some point, I came to realize that this "dead body" was the same man I had been talking to just hours earlier. I was devastated to watch him pass away right in front of my eyes, all while doctors and nurses were frantically attempting to drag him back to life.

Throughout those desperate lifesaving attempts, the man's heart never started, and he remained in a "flatline" state. He was as lifeless and as much of a corpse as he had been when I had walked into the room. While I watched this scene pan out, it became very clear to me that he had indeed been dead, even if the doctors had not yet officially declared him so, while they were pounding on his chest, administering drugs, and performing lifesaving maneuvers.

A slew of questions buzzed around my head. When in this hour or more of time had this man really died? Was there a clear distinguishing line between life and death, or had there been a transition between the two? Importantly, what had that man experienced during his transition between life and death? What had happened to his consciousness—*his selfhood*—and had he experienced any awareness while we were trying to resuscitate him?

In my then short experience of medicine, I had already heard of resuscitated patients who described being visually aware of what was going on around them as the doctors and nurses worked to save their lives. I wondered: *Had he experienced this? Did he retain consciousness, or was his selfhood annihilated when his heart stopped?* Eventually, after more than an hour of trying, the doctors stopped and declared him officially dead. This patient had now legally died, even though in reality he had died quite some time before, but my experience of meeting him and the questions his death prompted stayed with me and shaped my entire career.

At that time, I was also fascinated by what gives rise to our consciousness, our sense of self, and our unique inner universe of thoughts, all of which combine to make us distinctly human. Although I don't recall exactly what started this lifelong interest, a major contributing factor was my father, who suffered from a devastating brain illness that made him incapable of conscious responses during my childhood. On my visits to him, I would sit across from this vacant shell of the man I once knew and wonder what was left of him. Was his sense of consciousness, his selfhood and mind, still intact? From the outside perspective, he had turned into a "vegetable," yet none of us knew what was happening to him internally, or what his experience of this slow degeneration was like. There was no way to peer inside his brain and consciousness to try to decipher that unique inner world within. Instead, my family and I would hold his hand, look into his eyes, and wonder. We hoped that wherever he was, and whatever he was experiencing, he was not suffering. But science could not offer us any more than that.

Throughout medical school, I wondered: *How can brain cells give rise to thoughts?* They were cells just like any others, why should they produce thoughts? I wanted to know what made every single one of us so unique. At that time, I believed very strongly that the answers to my questions lay somewhere in the brain. I even contemplated formally pursuing a career in neurology, or psychiatry, but soon realized that neither would address these burning questions. That was because science did not yet have the answers.

I was also becoming more and more fascinated by the real life-and-death aspects of medicine. The events in the emergency room at Mount Sinai Hospital in New York brought all these blossoming scientific interests together: *What is consciousness? What happens to people between life and death? How can we restore life to people after their hearts have stopped?* And eventually directed me to pursue my current career path.

Now, as a professor of medicine, I have been privileged to research this subject throughout my career, while also witnessing its growth and evolution. I received my medical qualifications from the University of London, and I completed my graduate medical training at the University of Southampton, where I was also awarded a PhD in cell and molecular biology. Later, at Weill Cornell Medical Center in New York, I specialized in intensive (critical) care medicine. I am now writing about an adventure that I have taken through most of my adult life as a scientist and doctor whose expertise, besides intensive care medicine, lies in cardiac arrest resuscitation, the critically injured brain, and the brain and consciousness.

It seemed inconceivable then that just over a quarter of a century later I would be able to describe how our own discoveries, and those of other scientists, have addressed some of humanity's most profoundly significant yet seemingly unanswerable questions.

Although life and death remain a mystery, science is showing us that contrary to what some philosophers, doctors, and scientists have argued for centuries, neither biological nor mental processes end in an absolute sense with death. There is much more to be discovered, but science does, at a minimum, suggest that our consciousness and selfhood are not annihilated when we cross over into death and into the great unknown. In death, we may also come to discover that each of our actions, thoughts, and intentions in life, *from the most minute, to the most extreme*, do matter.

This scientific journey, much like the far more heavily funded, publicized, and well-known scientific exploration of space, also starts in the middle of the twentieth century. However, instead of touring the depths of space beyond our planet Earth, my goal is to take you through a tour of the far less well-known and exceedingly less well-funded, but vastly more significant, scientific discoveries into the deepest and furthest depths of the space that lies beyond life and into the newly discovered frontier of death. These innovations will allow us to peer inside people who traverse beyond death, to witness signatures of a unique hyperconscious lucid mental state emerge in their brains through bursts of activity. The *way* these brains reactivate, and the specific types of activity we record, together with analyses of many thousands of people's testimonies, will offer tantalizing glimpses into the reality of what we will all experience in death.

In the last few decades, scientists and researchers have made astonishing discoveries in terms of how the body dies, and what occurs when it does. In this book you will read about one of the most pivotal discoveries in this field, which challenges everything we've believed to date about the nature of death and when exactly we might consider an individual to be forever gone. Other studies are exploring consciousness in death. These studies will lead to my own work, uncovering the mysteries of what happens in the newly discovered liminal zone that lies beyond death, based on testimony from thousands of people who have been resuscitated. These experiments will have ramifications for all facets of our lives, both in the heat of a medical emergency and beyond.

We will hear from thousands of people who have walked into and returned from that liminal "grey zone" beyond death. These people, having had a recalled experience of death (or RED), return with a uniform, standardized series of experiences. It doesn't matter that they've never met or communicated with each other, or that they died in Mexico, Iran, Thailand, America, Denmark, or the United Kingdom. No matter how varied their lives, their encounters with death have a startling narrative unity.

This is not a book about religion or philosophy, and I will leave it to the reader to attach any kind of meaning to what these experiences mean and why they happen. However, there is no doubt that they happen. And there is a reason here for all of us to feel more peace and more acceptance about the suffering of our loved ones and our own inevitable demise. We may also learn more empathy and compassion for others, as a key component of the recalled experience of death is reliving your actions through the eyes of the people they affected.

We will start with the strange new world of the "brain in a bucket," and learn how everything we once thought about the supposedly quick degeneration of the brain after death is at best poorly understood, and at worst, wrong. If, as these experiments will demonstrate, our brains remain intact and revivable for hours after death, how, then, does this change our understanding of life and death and the experience of dying?

Section One

SCIENCE OF LIFE AND DEATH

Chapter 1

THE BRAIN IN THE BUCKET

In an unassuming boardroom tucked away on the sixth floor of an office building on the sprawling campus in a Washington, DC, suburb on the crisp morning of March 28, 2018, several men and women shuffled into the drab, beige room and took their seats. Microphones stood perched in front of each attendee's seat. Carafes of water, coffee cups, and plastic bottles littered the table, along with piles of paper and a few flickering laptops. A projector screen hung at the far end of the room, flanked by beige curtains and a lone American flag tucked into the far-right corner. The attendees looked around the room, down at the table, or up at the billowing blue screen, but rarely at one another. It could have been any office meeting in America. However, this was not just any office, this was not just any group, and this was not just any campus. This was a gathering of eighteen of the world's leading neuroscientists, ethicists, and legal minds at the National Institutes of Health (NIH). And they were all eager to hear discoveries that would open new frontiers for scientific advancement that seemed straight out of the world of science fiction. Either way, this meeting would change the world forever. It would also challenge one of the deepest and most fundamental issues underlying the fabric of our society—namely, the boundaries between life and death.

Dr. Nedan Sestan, a native Croatian, stood up and headed to the front of the room to take the podium. He apologized for the frog in his throat. Usually lively and excitable, Dr. Sestan approached his colleagues nervously and tentatively. After all, he had faith that his research was profound and important, but at the same time he knew that it could potentially shut his own work and lab down forever. In his forties, dressed in a check lavender button-down shirt, glasses, and with a full head of thick black hair, he looked young and energetic in contrast to some of his older and greying colleagues. You would hardly know he was a Yale professor, let alone a cutting-edge neuroscientist whose work has been honored by many prestigious scientific organizations and who is undoubtedly a future Nobel Prize winner.

He wasted no time. During a twenty-five-minute presentation, Dr. Sestan revealed the unthinkable. He and his team had successfully restored life to the brains of dead pigs. These were brains taken from the decapitated heads of pigs four hours after they had been killed for their meat in a pork processing plant. The brains were pumped with a solution containing artificial blood, together with a special cocktail of brain-protective and brain-stabilizing drugs through the brain's main blood vessels, using a specially designed life-support system called BrainEx.

To their total astonishment, over the course of six to ten hours (and eventually up to fourteen hours after the animals had died), Sestan and his team observed life return to the dead pig brains. In some cases, they had even managed to keep them alive and functioning for up to thirty-six hours. The brains pumped with the BrainEx solution showed pretty much all the functionality of a normal brain. There was no damage to the structure of the brain, its cells, or blood vessels, and blood flowed normally through them. There was normal brain metabolism, meaning biological and chemical activity. Electrical activity also emerged at cell connections, or synapses. This was astonishing.

Sestan explained that they had carried out this procedure already over three hundred times. This was certainly not a fluke. Admittedly, the pigs' brains had not shown the classical signs that scientists usually associate with being conscious and aware again, meaning their brain electrical waves had remained flat across the whole brain. He further clarified that this was to be expected, since they had all been given special drugs to avoid brain damage and to reduce the likelihood of a resurgence in global brain electrical activity, which is normally required for perception and awareness. They wanted to make sure the animals did not suffer.

Their results had been so astonishing that other scientists were convinced they were looking at healthy pig brains, not dead ones. Sestan acknowledged that restoring awareness and consciousness to these dead pig brains was a possibility. Crucially, he also disclosed the technique could, in principle, work on human brains and that these reactivated human brains could potentially be kept alive for a long time. This idea—that a mammalian brain could be dead but then hours later removed from the body, pumped with a solution, and reactivated back to consciousness—is astonishing. It also means that with some modifications it could be carried out as part of a strategy to revive a whole dead person. A reanimated brain is one thing in science fiction, another in life. How a human being might respond to a life lived completely within the confines of their brain, with no senses or connection to the world outside of its consciousness, is a question yet to be answered. However, it is also frightening to imagine that in the near future a dying individual might either volunteer or demand to have their brain harvested and preserved in this way.

SCIENTISTS HAD ALWAYS THOUGHT THAT WHEN A PERSON DIES, their brain cells become irreparably damaged within minutes due to oxygen deprivation. This had been the established scientific dogma for decades. Few had ever questioned it, and this idea has even fully seeped into popular culture. Now, Sestan and his team had turned this dogma on its head by showing that several hours after death, even though the brain is deprived of oxygen and nutrients, its cells do not die and degenerate quickly, and that life can be restored to the whole brain again. He acknowledged that what his team had achieved was both "unexpected" and "mind-boggling." With

BrainEx, they had created a system that had the power to restore function to the brains of mammals, potentially including humans, hours after death.

Sestan had just unlocked the secrets to reversing death. Moreover, if that was not enough, he had also opened the door to the scientific exploration of the postmortem period, and with it what happens to us all when we die.

I first heard about Dr. Sestan's research after I was contacted by *Nature* reporter Sara Reardon a year later, in 2019. She had sent me an embargoed copy of his completed study results and asked for my comments. I was left totally stunned and speechless. For at least a decade, I had tried to draw attention to the fact that our concept of life and death should be redefined. Death should no longer be viewed as a specific black-and-white moment. Instead, it should be understood as a medically treatable event for many hours after it has taken place. I had tried to draw attention to this in my book *Erasing Death: The Science That Is Rewriting the Boundaries Between Life and Death* (*The Lazarus Effect* in the UK), published in 2013, and in scientific publications and numerous lectures and interviews. Now, incredibly, we had arrived at this moment in time.

The audience that day at the NIH included Henry Greely, whom everyone calls Hank—a law professor at Stanford University and director of the Stanford Center for Law and the Biosciences and the Stanford Program in Neuroscience and Society. He is one of the world's foremost experts on the ethical, legal, and social issues arising from advances in genetics, neuroscience, and human stem cell research. There was also Nita Farahany, a law and philosophy professor at Duke University and a leading scholar on the ethical, legal, and social implications of emerging technologies and death. However, it was Jonathan Moreno, PhD, a professor at the University of Pennsylvania and member of the National Academy of Medicine, once dubbed the "quietly most interesting bioethicist of our time," who asked a question so powerful during the NIH meeting that it left Sestan literally speechless.

Moreno asked, "Forgive the conceptual grossness of this question. But now some hours later [after death] you have [turned a dead pig brain into] a comatose pig brain. This is how we can describe it at best."

It was a valid point. Sestan had turned not just one but many dead pig brains into what could at best be described as *alive brains in a coma*. Perhaps even brains that were conscious, with an active mind, yet in which the electrical signals normally associated with consciousness and awareness had not been detected because of the effects of certain specific blocking drugs.

To give perspective, there are plenty of people in hospitals all over the world who, like these pig brains, are in a coma with a flat brain electrical state (what the researchers had used to determine the pigs had not been conscious and aware). *They are not dead. They are in a deep coma, and with any coma, consciousness and awareness can potentially be restored with time.* However, with the pigs' brains, the issue was clear. To preserve the brains against damage, Sestan and his team had given special drugs that would block the transmission of electricity across the brains. It was not unusual, then, that those electrical signals were not detected.

The pig brains had been taken to death and then well beyond death and now they were alive again and in a coma. This was indeed a moral, ethical, and legal conundrum. What rights would the dead pig brains have after they were no longer dead? The seat of consciousness—*the mind*—in mammals, including in us humans, lies in the brain. Sestan had not revived the liver or kidney after death. He had revived the whole brain, and with that come all the ramifications about what it takes to be a person and to have a mind: mental activity, consciousness, awareness, and selfhood.

Sestan continued to sit still and blink for some time. He finally responded: "I don't want to do anything that is ethically questionable in my life. We did not get any signal [of consciousness on electrical monitoring of the brain] so that animal was not aware of anything. I am very confident of that."

He clarified that the animals had likely not suffered. That was a crucial point. In essence, the animal brains were similar to someone who is under powerful anesthesia and undergoing surgery without pain or awareness of the outside world. However, being pain-free and not having awareness of the outside world temporarily due to the effects of anesthetic drugs is not the same as not having consciousness at all—in the sense of being totally "soulless," that is, void and empty of personhood. Those pig brains did not show signs of being aware precisely because they had been given drugs that worked like powerful anesthetic medications in humans.

Sestan avoided answering the main question: What happens to animal brains that go from being dead—*as dead as possible, meaning decapitated dead*—for four hours or longer to then becoming alive again? No one pushed him for an answer either. That was probably because nobody else in that room could answer that question.

It takes enormous courage and audacity to even contemplate the idea of reviving dead brains, let alone carrying it out. There is no precedent in the history of science, and to most people it would have sounded more like something out of *Frankenstein* than real science. Yet here we were; Nenad Sestan had not only contemplated it, but he had also carried it out successfully.

As a developmental neuroscientist, he wanted to understand what makes us uniquely human and different from animals. This was what had initially set him off on the path that had led to this discovery. He explained: "This is what I have tried to understand for almost 20 years, since I started my own laboratory." He was convinced then, as he is now, that the secret to being human, the key to understanding who we are and what makes us all unique, relates to how the tens of billions of brain cells inside our brains are connected. It is not the size of the human brain, but the complex way those billions of cells, along with their long tracts, called *axons*, much like billions of long pieces of wire, are connected. It is worth delving a little deeper into how Sestan was able to accomplish this breakthrough with the pigs. His work here will directly affect many of us, in life as in death, and specifically anyone who requires an organ transplant in the near future. Most crucially, he more than any other scientist has cracked open and expanded the conversation about the nature of living and dying. In the process, he and his team have given further validity to the critical importance of research into what happens to the human mind and consciousness, including the millions of testimonies of a recalled experience of death, by providing concrete proof that the liminal grey zone beyond death is vastly more expansive than recognized in the past.

Dr. Sestan began his work in the early 2000s, an important time in the history of neuroscience. Just over ten years earlier, President George H. W. Bush had signed a presidential declaration designating the 1990s to be the "Decade of the Brain." And one year prior to the opening of his lab, on May 2, 2001, the results of a monumentally important scientific study called "Neural Progenitor Cells Recovered from Postmortem and Adult Tissue" had been published in *Nature*—one of the most prestigious scientific journals in the world. In it, Dr. Fred Gage and his team from the Salk Institute in California had taken pieces of postmortem human brains from people who had been lying dead for up to twenty-one hours in a mortuary. They had successfully dissected out brain cells from those dead human brains and had shown that they can grow again in petri dishes in the laboratory. Dr. Gage, the lead scientist, had remarked, "I find it remarkable that we all have pockets of cells in our brains that can grow and differentiate throughout our lives and even after death."

This *Nature* study had now fully confirmed Sestan's observations. So, his inquisitive mind started to drift to something new and without precedent. Might it be possible to create a complete map of all the major connections, known as the connectome, in the human brain? He likens this to how the internet is connected across the world, and how it has been able to transform the way information is shared, without changing the information itself. He wanted to map all the connections in the human brain and study how information flows across it. This would be audacious and costly. He would need to grow brain cells from dead people in the laboratory and study all their complicated connections. He explained, "Neither we, nor anyone else had ever done anything like this before. This was totally unknown." Understanding the connectome, he thought, was the key to understanding what makes us all human and different from, say, a chimpanzee.

The first step was to collect the brains required to study the connectome by working with a Wellcome Trust–supported brain bank in the United

Kingdom. Brain banks collect postmortem human brains from people who have donated them for research, and then distribute them to investigators all over the world. Sestan explained, "Usually, brain banks freeze donated human brains, before shipping them. One time, though, the brain bank placed a donated human brain on ice and shipped it to us without freezing it. They assumed the brain would have been cold enough to preserve it [against decay and damage from degeneration]. However, the courier service missed the flight to New York City. So, of course, the ice melted, and we received the brain very late, actually, 48 hours later."

Sestan felt awful. "Somebody had donated their brain in good faith. You do not want to [waste a] donation. You must be respectful and thoughtful. This was a very generous gift. However, after 48 hours we could not use it. We assumed it would have been far too damaged."

Instead of disposing of the brain, Sestan decided to find some other use for it. He had a graduate student, Andre Sousa, who needed to learn how to study the brain using a new research tool. He told Andre, "Why don't you practice on this brain?"

Two and a half weeks later, Andre ran over to Sestan in the laboratory and excitedly said, "I need to show you something." Sestan looked through a microscope at a petri dish and saw clusters of human brain cells growing healthily. "The cells looked as good as those we had grown from fresh pieces of brain obtained straight after surgery." Sestan was baffled. He knew the brain cells must have come from the postmortem brain that had been left for forty-eight hours and was supposed to have been damaged and degraded beyond repair. He did not believe what he was seeing.

He said, "At first, I honestly thought he was making a prank. I really thought that is what was happening." If this was not a tasteless, silly joke, then it was earthshattering. "The point was that the dead person's brain cells [taken from the brain that had been left for 48 hours] were growing naturally and even all the axons—the brain wires that link the cells together—were preserved." No one had ever done such a thing with a brain that had been left for that long after death. The dead person's brain cells were neither dead nor damaged. In fact, they continued to grow healthily for weeks.

Little did the courier company know how much their shipping mistake contributed to a huge scientific discovery. Just as Alexander Fleming accidently stumbled upon lifesaving penicillin growing and killing bacteria when he left an uncovered petri dish by a window, Sestan and his colleagues had unwittingly stumbled upon a scientific breakthrough that would change the course of medicine and science forever. This was his first Eureka moment. Although Sestan recognized the enormity of this discovery, he still had not imagined in his wildest dreams that it might be possible to revive a dead brain and make it function again. Another fortuitous and serendipitous event was needed for that.

SESTAN WAS STILL FOCUSED ON HOW HE COULD "TAKE A WHOLE [post-mortem] brain, keep it alive, and recover it." By "recover," he meant to take dead brains, remove large chunks out of the brain, and study the connections inside. This was a radical move away from how scientists had traditionally studied connections in small clumps of brain cells grown in a petri dish in the laboratory. Large chunks of brain were advantageous, as they would contain a much larger network of cells with preserved brain connections and longer tracts of axons—the wiring between brain cells. The study of the "brain connectome" excited Sestan, but he knew it would be impossible to conduct an experiment of this magnitude with brains from human cadavers.

The solution revealed itself one day while Sestan was out shopping at his local Costco hypermarket. He noticed there was a slaughterhouse nearby. Feeling courageous, Sestan walked up and knocked on the door. A large, grouchy, and somewhat tense-looking man appeared and looked at him with great suspicion. Sestan clearly did not look like a meat distributor. "I am a brain scientist from Yale, and I was wondering if you might have a spare brain?" Sestan asked. Totally bemused, the man became visibly more grouchy, tense, and suspicious. You would not expect a Yale professor standing at the entrance to your slaughterhouse. Maybe he was a food inspector in disguise. Sestan insisted, "No, no, no, trust me. This is for a scientific experiment." Eventually, after a lot of persuasion, the slaughterhouse worker finally loosened up. "We never use the pig brains. So, you can have as many as you want," he said. Sestan had found a solution to his problem. The pig brains were being discarded anyway. He could have as many postmortem pig brains as he needed and didn't need to have any animals killed specifically for this purpose either.

Until then, Sestan and his team had worked with very thin slices of brain or brain cells. There had not been any problem getting oxygen into them. Now they were facing a very different challenge. They needed to find a way to get oxygen into the large chunks of pig brain. However, because of the brain's size and delicate system of blood vessels, the sliced-up tissue struggled to circulate blood and, in turn, decomposed rapidly. "We would go a few times a week to the slaughterhouse. They would give us brains to work with. We tried for two years to make this work, but nothing worked."

One day Sestan had to attend a meeting to discuss an unrelated matter with Art Belanger, a pathologist and manager of the Yale University morgue. He looked over and saw something bizarre. There was a human brain hanging upside down in a bucket in a sink. As he watched carefully, he noticed a solution from a nearby plastic bottle dripping through special tubing into the main blood vessels of the brain and from there into the brain itself.

Belanger told him the rig was being used to "fix" and preserve the brain—a process called *plastination*. Sestan knew about fixing and preserving brains by immersing them in formaldehyde, but not by running a special solution—a method called *perfusion*—through the main blood vessels. "Trust me," Belanger told Sestan, "perfusion is much more effective." That was because, unlike immersion, perfusion used the existing network of blood vessels. It replicated the flow of blood through the whole brain. It would reach the deepest depths and all corners of the brain.

"That was a key turning point," Sestan told me. Now a light bulb had gone off in his head. Perhaps, instead of using a preservative solution as Art had done, he could use a different solution with oxygen and nutrients, and then pump it into the deepest depths of the brain through the main blood vessels.

Of course, it would not be easy, and even if he found a way, he still had to overcome the fact that it routinely takes upward of eight hours from the time of death to get a donated human brain to a researcher. Donated brains would routinely be left for hours without oxygen and nutrients, and all the blood inside their blood vessels would clot and no solution would run through them, meaning there could be no perfusion.

Sestan looked at Art and somewhat disappointedly said, "But this method could only work if it is started immediately after someone dies." Art, in turn, gleefully looked back at him and let him into a little-known trade secret. "No, actually the blood turns back into a liquid a while after death."

This was Sestan's second eureka moment. He was euphoric now. He realized that he had been thinking about the problem completely the wrong way. He had to work with an *entire* brain, and perfuse it, the way that Belanger was perfusing the brain in the bucket. He would have to switch from Art's fixative preservative solution to a nutrition-rich bloodlike solution that would carry oxygen into the brain in the same way that blood naturally perfuses the brain and carries oxygen and nutrients into every corner of the brain in every single one of us.

It would take some time to find the ideal machine to allow Sestan to explore the concept of perfusion. Nonetheless, Sestan eventually found a device capable of keeping kidneys, hearts, and livers alive outside the body for long stretches of time. The company, BioMEDInnovations (BMI), hoped the system, known as a CaVESWave, could help researchers study how to preserve isolated single organs designated for transplants.

Before they could even test out the machine's ability to enable perfusion of a solution into the whole brain, they would have to map out the architecture of the pig brain's blood circuitry with painstaking accuracy to determine how the arteries connected and what vessels to close off. They ran food coloring through the arteries to reveal the route the blood took to get deep inside every nook and cranny of each pig brain. This went on for months. After many pig brains, the team had mapped out their course and had a plan.

As soon as the BMI machine was delivered, Sestan and his team started to modify nearly every aspect of it, by tailoring it to the unique needs of the experiment. They named the system BrainEx and jokingly nicknamed it "brain in a bucket." The astonishing power and potential of BrainEx was undeniable.

Sestan and his team turned the machine on for the first time while they worked on creating the perfect perfusion solution for the brain. They decided to use an artificial bloodlike substance called a hemoglobin blood oxygen carrier (HBOC). They then added at least nine different types of brain protective agents to enhance blood flow and protect brain cells. They needed these to fight against the dreaded "reperfusion," or "reoxygenation injury," that kills brain cells rapidly once oxygen is reintroduced to the brain after it has been deprived of it. Chemicals were also added to allow the scientists to track the flow of this solution throughout the brain using ultrasound.

The time came: the scientists pumped a wine bottle's worth of the dark, scarlet red HBOC liquid, mixed with the cocktail of protective drugs, into the brain for one hour. Because blood does not circulate through arteries in a uniform fashion, the machine needed to mimic its natural rhythm. They made another key mechanical modification to the system: an automated pulse generator device that replicates the pulse of a heartbeat. After a series of critically important tweaks to the mechanisms of the machine, they began to observe the brain tissue had a pinkish-grey hue resembling that of a living brain.

Placing the first pig brain into the machine marked the beginning of an experiment that would jolt the door open to a new era in science and unravel a confounding set of questions for scientists and bioethicists to grapple with around the world. By the time the experiment started, many hours had passed since the pigs had been killed. Until now, scientists assumed that when the blood supply is cut off, the brain declines and degrades quickly through a series of steps that lead to brain cell death. They thought those changes were irreversible unless blood was quickly restored. It had not been truly tested before. How would they know where the line between possible and impossible was drawn unless they tested it? They poked and prodded at those possibilities as they increased the length of perfusion time from one hour to two and from two to three. With each subsequent hour, more brain cells were restored.

For the first time in human history, using a machine originally devised for maintaining single organs outside of the body for transplantation, *researchers now had the capability of studying the entire brain outside of the body after death*—allowing them to map cells and their connections with a level of precision that had never before been possible. Sestan recalls his astonishment as he hunched over the microscope. The brain and its underlying cells were biologically active again. Brain function had been restored. *Oh my God*, he thought to himself.

Sestan was acutely aware that the brains might regain conscious awareness. Imagine, for a moment, if the pig had become conscious but had no eyes to see, no snout to squeal, and no body, but was nonetheless aware of being disembodied. It is the stuff of science fiction or our worst nightmares, but nonetheless, it was a real possibility. They needed to ensure that the brain, once dead and reanimated, would remain without conscious awareness, to avoid an indescribably horrifying situation for the animal. The longer the experiment ran, the more likely it would have been that awareness and mental activity could emerge. Whether he liked it or not, he had inadvertently opened up a new frontier of science: the age-old philosophical territory of what happens to us all—meaning mind, consciousness, and selfhood in living beings—after death.

He needed a machine that would help him find out. He turned to a bispectral index (BIS) monitor, a device that anesthesiologists use to determine the depth of a person's consciousness and awareness while under a general anesthetic drug during surgery. It measures brain electrical signals and converts them into a number. The results are expressed on a scale of zero to 100. The higher the number, the greater the electrical activity that corresponds with full consciousness and awareness. The BIS monitor on one of the pig brains measured 10, not zero, which was about the level of someone in a deep coma. Sestan unplugged the machine. This was frightening. Had the pig brain been conscious after death? This was uncharted territory.

It was time for bioethicists to weigh in before Sestan went any further. He immediately wrote an email to the director of the Interdisciplinary Center for Bioethics at Yale and a second one to the Neuroethics Working Group of the NIH BRAIN Initiative that had funded his research.

After discussions with scientists from the NIH, it was decided that since the BIS monitor is designed for use on a human skull, it may have picked up a signal in error when placed on a pig brain that was outside the skull. Therefore, the recommendation was to use a different type of brain electrical monitoring system called electrocorticography, or ECoG for short. This would work better as it was designed to be placed directly on the brain during brain surgery, rather than on the bony skull. Sestan was also advised to use strong anesthetic drugs, and to reduce the brain temperature rapidly to swiftly diminish any visible signs of consciousness, should any be detected. Using ECoG, and with those special brain blockers on board in the perfusion solution, Sestan and his team never detected electrical signs of conscious awareness again.

On that fateful day, March 28, 2018, Sestan was meeting with the NIH in what he thought would be a confidential setting. Soon after, reporters at the *MIT Technology Review* got their hands on a video of the meeting and published an article on the findings. Within hours, media outlets around the world were running stories with clickbait titles about "severed heads" and "Frankenstein-style research."

This kind of sensationalism bothered Sestan, but he continued his work. He realized that if they could perfuse the brain using a bloodlike substance with oxygen and nutrients, they could potentially restore life to a dead brain. Sestan was right when he had said, "We didn't plan to restart life in dead brains. This was a really out-of-the-box project for us." Now, he was on the cusp of doing so. Of course, conducting this research project in human brains was out of the question. Nonetheless, the experiments he was about to start would no doubt set the religious, bioethics, scientific, and medical communities into a frenzy.

One year after his NIH meeting, Sestan's study was formally published in the world-renowned medical journal *Nature* on April 17, 2019, under the title "Restoration of Brain Circulation and Cellular Functions Hours Post-mortem." It was remarkable. It provided a step-by-step account of how he and his team had restored life to the brains of thirty-two decapitated pigs, reviving them for up to a fourteen-hour period postmortem.

Immediately, news headlines spread all over the world like wildfire. "'Partly Alive': Scientists Revive Cells in Brains from Dead Pigs," was the *New York Times* headline. *The Guardian* reported, "Scientists Restore Some Function in the Brains of Dead Pigs Brain Cells—Four Hours After Death." "Basic Cellular Activity Pigs Brains Hours After Death," enthused the *World Economic Forum*. Another English-language headline was "Pig Brains Partly Revived by Scientists Hours After Death." Similar headlines appeared across all major media outlets and in every language and country.

While Sestan's discovery was the talk of the town, all the headlines were somewhat underwhelming, given what had just been discovered. The media were reporting that the brains were *not* alive and that some vague semblance of "limited cellular" activity or "partial" brain function had been restored. These reports were grossly inaccurate and frankly misleading. The newspaper reports underplayed the importance of the discovery. Sestan had just—almost singlehandedly—ended the assumption that life and death exist as binary, easily separable entities and that once you leave the safe shores of life, there is no turning around. What does death mean if it can be reversed and life restored again after an hour? Or two hours? Or four, or eight, or ten? It's one thing to approach this question as a researcher and scientist, but what does it mean to each and every single one of us as human beings, and how will it impact a lawyer, ethicist, theologian, or philosopher? Importantly, what will those people who are revived back to life after crossing beyond the biological bounds of death have to say about their experience in this newly discovered grey zone of death?

Professor Steve Waxman, the senior neurologist at Yale University, advised Sestan that he needed to understand how "this would have a profound effect on people's awareness of death . . . [because] people think about this as a binary event—[meaning] you are either dead or alive." He told Sestan, "You are not ready [to deal what is going to come], because you have not been thinking about that enough and you have to be very careful what you say."

To give perspective, it was as if a scientist had just discovered a drug to cure all forms of cancer—even end-stage cancer—within a few hours of administration, yet the media were reporting that the drug had some "limited cellular" activity against cancer cells, without indicating its true potential to cure.

Instead, there were at times outlandish and frankly incorrect and unfounded assumptions about the absence of life and consciousness. This was exemplified by *The Guardian*, which wrote: "The brains were neither alive nor possessed consciousness . . . Some neurons even started firing. However, there were no signs of the coordinated, brain-wide electrical activity indicating sentience. The team had anesthetics on hand in case brain activity did indicate consciousness. It never did."

The study had demonstrated that dead mammalian brains can be fully (not just partially) revived and that the BrainEx technology could be adopted to fully revive dead human beings. Sestan and his team had restored life completely—*by that, I mean all aspects of measured brain function*—to decapitated pigs' brains up to fourteen hours after death. The fact that consciousness—sentience—was not detected was simply because the BrainEx solution included a drug that acted like an anesthetic and blocked electrical transmission. Since they were using transmission of electricity across the brain as a sign of consciousness, then of course they did not observe it. They very likely would have if they had simply removed the drug from the solution.

This was an enormous discovery. Could there be any news bigger than the formula to how to reverse death? I was baffled. How could something so big be so widely misreported and downplayed? Even most experts, who were quoted by the media, seemed to ignore the huge ramifications that this and other recent scientific breakthroughs were having on our understanding of what happens when we die and the ability to restore life after death.

Reading between the lines, it became clear that the downplaying of the study findings was actually born out of a collaborative effort led by the researchers, the NIH staff, and Yale University staff themselves. One key to understanding what happened behind the scenes came from a conversation I had with science writer Jules Asher, who worked on the official NIH press release and retired just a few months later from the NIH's Office of Science Policy, Planning, and Communications after forty-five years.

When I spoke with Asher in the summer of 2020, he explained that the public relations teams at the NIH and Yale University had purposefully decided to work together to greatly soften the study's findings for the public. He said, "We just didn't know what to do with these findings. So we decided to use language that would be vague so as to downplay the findings, especially those that dealt with issues of life and death. We just didn't know enough about this. So we decided to be cautious." They had written the press release with hypervigilance, weaving words together with caution so as not to elicit mass frenzy.

This seemed like how political parties and governments use language to spin things in order to hide facts from the public and media. This was certainly not what we expect from our scientists, our scientific funding bodies, or our universities. The careful and collective effort had worked, by steering the world media away from the real findings of the study.

This was exemplified by a very deliberate use of scientifically sounding, yet illogical and contradictory language to detract away from any discussion about the restoration of life to dead pig brains. The reason for the media erroneously reporting that life had not been restored was that Zvonimir Vrselja, a researcher working with Sestan, had been quoted as saying: "Clinically defined, this is not a living brain, but it is a cellularly active brain."

How can anyone say that a cellularly active (meaning *biologically* active) brain is not alive? Isn't biological activity the definition of life? This was inconsistent with any clinical or scientific reality. Yet no one seemed to question it. By that logic, then, every single one of us with biologically active brains must redefine ourselves as not having a "living brain" but a "cellularly active brain." The selective quote had worked incredibly well, and it had duped everyone, as intended.

Sestan himself and most of the people from the NIH involved with the press release had been together on the NIH grounds a year prior to the study's publication and had discussed how the pig brains had been alive but in a coma. That was Dr. Moreno's point. Sestan had also alluded to the fact that if they had simply removed the blocking drugs and had waited a little bit longer, they could very well have seen full electrical activity consistent with consciousness, awareness, and thought processes emerge in the pig brains. This was why he had acknowledged that restoring awareness and consciousness—*activity of the mind*—to these dead pig brains was a possibility.

The study grapples with fundamental questions of what it means to live and die. However, the press release minimized the results, and they were dimmed and even squashed by the people who had funded and carried out the work itself. Only those with the scientific acumen to see through the charade could truly decipher through the study findings, with mouths agape, in awe of the implications for humankind. The scientists and the NIH staff, who had funded the study, were ill prepared to deal with the ramifications. They needed time to get to grips with the enormity of what had happened. Their discovery had been too big, too profound, too quick, and too unexpected. There had simply not been enough time to wrap their own heads around what they had discovered. Therefore, they felt it necessary to be prudent and extremely cautious.

The public rightly expect thoughtful answers from scientists. Sestan acknowledged that he himself was taken aback by the results of the study. "I myself did not deeply understand it, and that is the key." He had stumbled upon something far bigger than he could have ever imagined in his wildest dreams.

There had been no malice in the soft-pedaling of the results, simply extreme caution. Behind the convoluted language and media spin, the fact remained that the technology and science behind BrainEx would be the key to restoring life after death.

SOON AFTER, THE DOMINOES KEPT ON FALLING. FIRST INSPIRED BY Sestan's breakthrough, another group of scientists went on to push the envelope of death, too, and published the results of their efforts in May 2022. Through a collaboration between Scripps Research and the University of Utah, scientists led by Dr. Fans Vinberg and Dr. Anne Hanneken had taken the eyes of dead people, five hours after death, and then restored nerve and light-sensing functions to those human eyes. This would have seemed unthinkable before Sestan's work. Yet it was now contemplated and even successfully accomplished in a short period of time.

Then it was Nenad Sestan's turn to shock the world again, in August 2022, through an experiment that would provide a huge step toward reviving dead humans in the near future. This time, he and his team studied the ability to revive completely dead pigs, not just the pig brain. They introduced us to the power of OrganEx, a new machine and procedure derived from BrainEx that could revive all the organs simultaneously in dead pigs, starting an hour after death. They compared dead pigs treated with OrganEx with another group of dead pigs placed on extracorporeal membrane oxygenation (ECMO). This machine is only available in select high-caliber hospitals today and resembles a heart-lung bypass machine. It pumps blood around the body, supplies oxygen, and removes carbon dioxide. It is considered far superior to standard methods of resuscitation, such as CPR. In short, they compared OrganEx head-to-head with the best of the best, the Rolls Royce of resuscitation equipment.

What was the result? Starting one hour after the animals had been killed and over six hours (the length of the experiment), OrganEx won easily by revitalizing all the major organs, including the brain, heart, lungs, liver, kidneys, and pancreas, *hence reversing processes of cellular death that we had never seen in science.* With the pigs treated with ECMO, the scientists found serious signs of damage in all the main organs. However, OrganEx-treated pigs passed various tests that showed the return of normal bodily functions: glucose, blood clotting, kidney function, and heart activity. With OrganEx, there was also more repair and recovery. Importantly, researchers found evidence that brain cells had recovered, too, as they were able to absorb blood sugar and oxygen and produce carbon dioxide, all key signals that they were healthy and functioning.

For now, officially, Sestan and his team, the NIH, and Yale University again chose to steer away from controversies as far as possible and downplayed their findings. They avoided mentioning the obvious application of OrganEx in reviving dead people, as well as its implications for blurring the boundaries between life and death. The official line was that OrganEx can transform countless lives—not by bringing back dead people—but by expanding the organ donor pool to numbers that we've never seen before in medicine. This technology was proposed to help doctors revive organs in an individual who is dead by recreating an artificial circulation and perfusion, which in turn can transform a deceased person into an organ donor, to enable them to give life to others. *The obvious point, the elephant in the room, is that if you revive organs in a dead person, then that person becomes alive again.*

The studies by Yale University scientists demonstrate that the line between what we may consider life and death is far more blurred than we might care to imagine. As a scientist I am concerned with how this knowledge will help us develop new treatments to revive and restore life to people well beyond death. I am also concerned with what happens to us all: human consciousness, the self, and the very human experience of going through death. The pigs' brain study confirmed to me something that I myself have been working on for more than a decade, proof that death is a continuum, rather than a binary event. Even though these pigs cannot share their experience of life, dying, and death, the research on their brains was crucial to validating my own work on what happens to the human mind and consciousness in death.

It is just a matter of time before OrganEx is used to restore life to a dead person. In fact, scientists are actively working on this and we hope to join them. I realize this story may sound like science fiction, but it is not. It is absolute science reality. If the idea of having yourself resuscitated as a "brain in a box" sounds vaguely horrifying, or at the very least, ghoulish, the idea of your whole body being resuscitated by OrganEx sounds hopeful, astonishing, and life-affirming.

The liminal space between life, death, and beyond is widening. And in the future will only get wider—as people who have died will be revived by technology like OrganEx. This means science can no longer ignore the need to conduct empirical and unbiased research into what happens to human consciousness beyond death. These people, who have spent hours in the grey zone beyond death, and who will be able to tell us about their experiences in death, may be the key to understanding what happens to us all after our hearts stop beating and we die. The insights, observations, and reflections they have to share may profoundly change how we view our selves, and how we choose to spend our brief sojourn in this world. It may be that it is only in dying that we truly learn how to live.

Chapter 2

A NEW SCIENTIFIC FRONTIER

IT IS IN OUR NATURE TO DENY DEATH. WHAT NENAD SESTAN HAS managed to achieve represents the culmination of work with roots in ancient times. Our ancestors tried and failed to revive stopped hearts and bring breath back to lifeless bodies: physicians in ancient Greece warmed dead bodies with fresh, steaming excrement or hot ash on the belief that life is associated with warmth, and death with cold. In other ancient cultures, physicians whipped dead bodies or blew smoke in them to try to restore movement and hence life. Some others tried tickling dead people's throats with feathers, again to stimulate movement as a sign of "life." By the fifteenth and sixteenth centuries, Europeans had started to use fireplace bellows to blow air into the lungs of drowning victims, and by the eighteenth century, dead people were being rolled on barrels, or placed on a trotting horse to move the chest up and down, in the hope of restoring life. None of these techniques worked and the dead stayed dead.

Medicine, as with much of science, underwent a revolution in the twentieth century and things started to radically change, particularly after the rapid advancements in science and technology in the 1950s. This was when the structure of the DNA molecule was discovered; the first satellite was sent to space; modern computers were made; and of course, modern ventilators and life-support systems were developed, which prevented people with life-threatening conditions from deteriorating to a point where their hearts would stop beating and they would die.

This laid the foundations for modern intensive care medicine, which has saved many hundreds of millions of lives. But with the discoveries of modern medicine came many intriguing ethical and moral dilemmas, which started to challenge some of the most strongly held social concepts and conventions, perhaps none more so than our understanding of life and death itself.

Throughout time, a stopped heart signified the complete and permanent end of life. The moment the heart stops is a black-and-white, binary event: your heart is either beating or it is not. Our laws and traditions are based on this. For example, while a "living" person has many rights, a "dead" person does not. The modern history of resuscitation science—the ability to reverse death—started at the beginning of the twentieth century, initially through primitive separate efforts by scientists to develop drugs to bolster the blood pressure, breathing devices, and electrical shock therapy for the heart. In 1959, William Kouwenhoven, James Jude, and Guy Knickerbocker, at Johns Hopkins Hospital in Baltimore in the United States, put these things together and finally discovered a method to restart the heart in people who would otherwise be permanently dead.

They tested their new discovery for the first time on a thirty-five-year-old woman who had unexpectedly died a few minutes earlier in the operating room. Remarkably, they managed to restore life back to her. They announced their breakthrough system of "cardiopulmonary resuscitation" (CPR) to the world and started a worldwide movement that would go on to save many millions of people's lives. However, their discovery would also raise a major question: If someone who would otherwise be permanently dead can be brought back to life again, what does this mean for our understanding of life and death, and in particular the idea of the absolute permanency of death? Their work had led to a blurring of the boundaries between life and death.

Throughout the mid- to late twentieth century, research into this new field of resuscitation science and intensive care medicine progressed. The Russian scientist Vladimir Negovsky became one of the fathers of "reanimatology" (a term he invented, meaning to reanimate, bring back to life/breath), or what we now call "resuscitation." His American compatriot was Dr. Peter Safar, who formed the first center for resuscitation science at the University of Pittsburgh. Safar and Negovsky forged an enduring friendship and professional collaboration. They discovered much more about the early science of resuscitation beyond what those pioneers had achieved at Johns Hopkins Hospital, including important advances in modern CPR methods, research into the brain, and the multidisciplinary specialty of critical care, or intensive care medicine.

While the Space Race and the race to find ways to "reanimate" the dead were going on at that time between the Soviets and the Americans, there was one major difference between the two. Resuscitation science wasn't as glamorous as, say, a "Space Race." The pioneering Gemini missions of the 1960s carried only two astronauts each. They orbited the Earth for periods ranging from five hours to fourteen days. The program consisted of ten crewed launches, two uncrewed launches, and seven target vehicles, at a total cost of approximately 1.3 billion dollars. Not since the building of the Panama Canal in peacetime or the Manhattan Project during war had the United States spent so much, so fast, on one specific project. Today, although the National Institutes of Health spends billions of dollars every year on medical research, unlike with issues such as opioid addiction, cancer, and Alzheimer's disease, it has no dedicated program to fund the study of resuscitation, even though every single one of us will suffer a cardiac arrest, as this is the only universal medical condition.

Around the same time that CPR was discovered in the United States, across the Atlantic Ocean, two eminent French neurologists, Pierre Mollaret and Maurice Goulon, working at the prestigious Hospital Claude-Bernard in Paris, announced a bizarre and extraordinary finding to the world. At that time, the new field of intensive care medicine was still very much in its infancy. Doctors around the world were starting to use ventilators to keep people alive who would normally have died when their hearts stopped and their breathing ceased after devastating brain injuries from car accidents and gunshot wounds.

Nevertheless, Mollaret and Goulon, who had examined such cases in their hospital, noticed something strange. Although these people were all connected to ventilators and consequently were getting artificial breaths and had a heartbeat and so were technically alive, many of them just did not seem alive. In fact, they resembled an empty shell, a husk of a person. This was a new phenomenon resulting from the use of life support and ventilators. Those patients remained in a very deep coma—*the deepest coma these doctors had ever witnessed*—with no signs of consciousness. Within a few days or weeks, their hearts would stop beating, they would stop breathing, and they would be declared dead in the conventional manner. Curious to understand more, the physicians performed a postmortem examination on these strange cases.

To their astonishment, they discovered that a thick liquid, *a sort of sludge*, had replaced parts of the brains of some of these people. Importantly, they realized that those people's brains had died some time ago and some sections had gradually disintegrated and liquefied in their skulls before their hearts had stopped. Unable to fathom what they were witnessing and without a precedent, they called this condition *coma depasse*, literally "beyond the deepest coma."

These two French physicians grappled with a hard dilemma and a very challenging ethical question: Without a brain, *the seat of the human soul, personhood, and consciousness*, could these people have really been alive, or had they perhaps been dead for days and weeks already despite having a heartbeat? Surely, without a brain, they could not have been considered alive. If so, when had they actually died and what had happened to their brain and their consciousness—their real self—while they had been attached to life-support systems? Unlike a heart stopping, there was no absolute binary, black-and-white, biological moment that they could use to draw a line to say life had ended and death had started. This represented an even more powerful blurring of the lines between life and death than the discovery of CPR. With these cases, there were clearly shades of grey, but not a clear black-and-white line between the two.

Other doctors also started to make these same observations in other parts of the world, too. For many doctors, ethically it did not seem right to wait days or weeks for the hearts of people with *coma depasse* to stop beating, just to meet traditional social concepts of death. They argued that if, indeed, the brain had died already because of a devastating injury, then irrespective of what was happening with the heartbeat, the person must also be considered "soulless" and dead.

Dr. Stephan Mayer, a professor of neurology and neurological intensive care at Westchester Medical Center in New York, explained, "It was recognized that after catastrophic brain injuries, trauma, bleeding into your brain, these diseases could effectively kill off the brain permanently to the point that there's zero brain function, and no blood flowing to the brain at all. Yet, the heart is still beating because the ventilator is sending the oxygen into the lungs." That is because people have lost all brain functioning and are not able to breathe. So, if you turn the ventilator off, they can't initiate a breath because the injury to their brain extends all the way down to the brain stem, the most primitive part of the brain, which controls all the things that we don't think about, like our bowels contracting and our hearts beating and our lungs initiating breath. But their heart is still beating, and other biologic functions are still going on, yet they are unable to breathe without a ventilator.

Throughout the 1960s, there were ever-increasing calls within the medical community to allow such individuals to be legally declared dead, without having to wait for the heart to stop. Eventually, in 1968, formal criteria were written recognizing and endorsing "brain death" as a new entity and a valid way to die. This was gradually adopted, albeit with some variations and with some delays, by different countries and in different jurisdictions around the world.

Today, it is estimated that around 98 percent of people are declared dead based on the time that their heart had stopped beating (cardiac arrest). This is called death by cardiopulmonary criteria. The other 2 percent are legally declared dead after devastating brain injuries, using "brain death" criteria, without waiting for their hearts to stop beating. These advances led some people to mistakenly believe that brain death has somehow superseded cardiac death. That is clearly not the case. It is simply another legal mode to allow doctors to declare people with devastating brain injuries dead without the need to wait for the heart to stop.

Progressively, throughout the twentieth century, doctors also started to understand that time was a major factor for the brain to become irreversibly and irretrievably degraded—in other words, damaged—and "die," whether that be after a devastating brain bleeding disorder or a massive stroke. Nonetheless, the assumption remained that if someone suffers oxygen deprivation, as happens after the heart stops and a person dies, then the brain will be permanently damaged within minutes.

Now, the twenty-first-century work of scientists, exemplified by the research of Dr. Nenad Sestan, has shown that brain cells and the brain do not degenerate and become irreversibly and irretrievably damaged for hours or days after we all die, just as they don't after other brain injuries. In other words, it has been discovered that the brain goes into a sort of hibernation state after people die, retaining the potential for repair so that it can be brought back to life again. Ironically, the reason why people end up with brain damage after oxygen deprivation is that when the heart is restarted, the oxygen that flows back again becomes toxic. This is what causes the cells to die quickly, much more than in the period in which they were deprived of oxygen with death itself. As mentioned, scientists call this reperfusion, or reoxygenation injury, and it is this that Dr. Sestan had managed to overcome using his novel brain-protective cocktail. That is why the dead pig brains could be revived. Either way, the discovery that cells don't die quickly after death is the third and most significant challenge—after the discovery of CPR and the emergence of the legal concept of brain death—to our longstanding societal but outdated concept of a binary separation between life and death.

Until now, doctors and society dealt with the dissonance between modern scientific discoveries and traditional social concepts regarding life and death by taking the easy route. For example, when CPR was discovered, doctors simply used different terms to maintain a differentiation between the two. They referred to stopping of the heart as "cardiac arrest" when they performed CPR, and "death" when attempts at CPR were not made or were stopped. This semantic distinction was related to neither the biology of life and death nor the process that happens in the body when someone dies. Yet it did help avoid public confusion and relieved any dissonance in society and avoided the need to answer difficult questions about the blurring of the boundaries between life and death. Today, if someone has requested not to receive CPR (or other resuscitation attempts), doctors declare that person "dead" when the heart stops. If they receive CPR, which often only lasts about twenty to thirty minutes, they call it "cardiac arrest." Then, if the heart cannot be started, they immediately call it "death." Biologically, this is all still the same process—i.e., the heart stopping and whole-body oxygen deprivation. Nothing is that different in the body in those twenty to thirty minutes. The only difference is a semantic one.

To give some perspective, of the roughly 3.4 million people who die every year in the United States, 500,000 to 1,000,000 (around 15 percent to 30 percent) undergo CPR attempts. Typically, they are initially labeled "cardiac arrest," and when CPR fails, they are declared "dead." The remaining 2.4 million to 2.9 million (70 percent to 85 percent) are simply declared dead immediately after their hearts stop. These are people who never wanted to be revived, who were not around people who could or knew how to save them, and/or who were very old or had terminal illnesses.

Even after the concept of "brain death" was discovered, doctors still tried to relieve the dissonance between the social notion of life and death as a binary event and the scientific reality of a grey zone by drawing a line to distinguish "death" from "life." Even though there is no absolute biological moment that distinguishes life from death in those people on ventilators with catastrophic brain injuries, their "time of death" was based on the time when the doctor had managed to conduct tests showing lack of brain function. So, in this way, the black-and-white societal line was maintained. If there were delays in performing those tests for instance, a doctor could be attending to another patient and not able to reach the person whose brain had died, or families were not ready for testing to take place for days—the "dead" person would still legally be considered "alive" until the doctor had completed the necessary tests to call it. Even though this is not biology or medicine, it was a way to adhere to an arbitrary and rigid social convention.

Of course, as already discussed, injured brains do not behave in a binary manner. A severely injured brain undergoes progressive damage and degradation over time. Importantly, well before the damage has become irreversible and irretrievable from a biological perspective—in other words, well before the brain has "died"—the brain loses function. The question of how long doctors should wait to be certain that the loss of function they test for reflects permanent, irreversible, and irretrievable damage—hence degeneration and death—is not clear. This is because it does not occur at an absolute specific moment in time, and that is why in different parts of the world there are nuances and different laws for when brain death testing should be carried out. These vary depending on the nature of the injury, typically from twelve to seventy-two hours. That means the same person with the same devastating brain injury, such as a brain bleed or massive stroke, could be declared "dead" in one jurisdiction or country and "alive" in another. This further highlights the fact that trying to draw an absolute line between life and death based on a biological point is arbitrary. There are also cases of people, including children, who have been declared dead using brain death criteria. However, their families had not agreed with the declaration of "death" and instead had insisted on maintaining them on ventilators while they still had a heartbeat. Consequently, some had stayed in that state for years and, later, had gone on to complete puberty. This means that not all parts of their brains (for example, the areas that deal with puberty) had died, and our idea of a clear separation between life and death, or that the whole brain always dies in people who are legally declared brain-dead, is oversimplified.

In view of the ever-increasing scientific challenges to the long-standing conventional social notions of a black-and-white separation between life and death, in 2021, the Executive Committee of the Uniform Law Commission in the United States unanimously voted to create a drafting committee to update, refine, and modify the legal definition and consideration of death. This modification is being sought only forty years after the 1981 Uniform Determination of Death Act (UDDA), which has been the legal statute for determination of death in the United States since its adoption.

Considering there have been many similar statutes and declarations, albeit with some differences across the world, how is it that something so fundamental and seemingly universal and simple as "life" and "death" needs to be reviewed and revised every forty years or so, and why are there differences in the declaration of brain death across different countries? Clearly, at the very least, these discoveries and events show how our understanding of the fundamental issue of life and death and what happens when we die is not a simple black-and-white matter.

I KNOW THAT WHAT I AM SAYING SOUNDS COUNTERINTUITIVE AND hard to fathom, but it shouldn't be. Imagine you are flying across different time zones, say, from North America to Europe or Asia. You may start out in the daytime zone where there is absolute light, but then gradually transition through less and less light and more and more darkness over several hours. Eventually, there is no light at all, meaning absolute darkness. We all know there is clearly a daytime and a nighttime. However, there is no clear line that can be drawn during a flight to create a binary concept of darkness and light. It is a continuum over many hours. In the same way, there is no absolute clear biological line that can be drawn between life and death from a scientific perspective, at least not for a number of hours, if not longer, into the postmortem period.

We can no longer try to simply get around this by using new terms, or even claiming that people who have died were not dead, because the scientific dissonance does not fit with our social concept of death. If we truly believed such a thing, we would have to also believe that almost everyone who has been declared dead after their hearts had stopped from the beginning of time until now, including all those who have died of heart attacks, infections, cancers, war, or famine, or during the various pandemics, had not really died. Clearly, they were all dead. Instead, we need to face reality. Science is logical and clear: our social concepts are arbitrary and may not always be correct.

Science has shown that what we call death is not an end, but actually is *a medical condition* or disorder that is amenable to treatments for many hours. It can be likened to a *transient ischemic attack* (TIA), meaning a transient and temporary period of oxygen deprivation to the brain. In the case of death, this affects the whole body, not just the brain. The term *TIA* is traditionally used by doctors to refer to people who develop the signs and symptoms of a stroke in the brain, such as one-sided weakness. However, what distinguishes a TIA from an actual stroke is that all the deficits with a TIA are relieved within twenty-four hours. The temporariness of a TIA, unlike a stroke, which is permanent, is due to a temporary deprivation of blood (and hence oxygen and nutrients) to the brain, which leads to loss of function (e.g., one-sided weakness) like a stroke. However, if this extends beyond a certain point, it leads to irreversible damage and degradation, meaning a stroke, which represents permanent loss of function.

Importantly, this explains why what we call "death" should be viewed as a medical condition that can and should be treated and potentially reversed even many hours after it has occurred. When we label this period of whole-body oxygen deprivation as "death," or a "permanent end," we do irreparable harm. Why? Because when society labels something as the "end," then naturally no one thinks to try to go beyond and find treatments. This is what made Dr. Sestan's work even more remarkable. He singlehandedly defied convention and traversed well beyond what everyone considered the end and showed that it is not. This has opened up a new field of discovery and potential new therapies for people who would be dead.

We now need to adopt and refine the techniques developed by Dr. Sestan and apply them to the millions of otherwise healthy humans—athletes, people with heart attacks, and car accident victims—who will die, *but who could live with a good quality of life* if we treat and reverse their death. We also need to face up to and address the many ethical questions that have come from this discovery.

So, let us deal with the elephant in the room. Often, people, including many scientists and doctors, say if your heart has stopped and you are alive again, then you could not have died because "death" is irreversible and permanent by definition. People have even created terms, such as *clinical death*, to distinguish between people who have traversed death and returned from people with "real" death, in their own minds. However, this is just a play with words. It has nothing to do with biological and scientific reality.

Surely, nobody would try to argue that those pigs who were revived by Dr. Sestan were not dead, or that a person who is decapitated, like those pigs, is not dead, or that someone who dies of massive bleeding after a car accident is not dead, or that people who have been declared dead and taken to a mortuary are not dead. *They are all as dead as can be.* However, science has shown that in each case—although they are dead and their bodies and internal organs are not working, and they show no signs of life—the cells in those nonfunctioning organs, including in the brain, will not degrade and become fully damaged and "die" *for many hours.* That is because of a long period of something resembling a hibernation state after death. This is the science that enabled Dr. Sestan to achieve this remarkable and truly monumental feat of reviving dead mammalian brains.

Of course, science still agrees that an irreversible loss of life is "death." Nevertheless, the major revolutionary discovery is that the process of cell damage and death in the body, including in the brain, generally starts *after* a person has died and does not become absolute for many hours or even possibly days after death. The transition from life to what we call death, like everything else in biology, follows a continuum. No biological process is a binary, singular, black-and-white moment. Why should this be any different? Yet for social reasons, we are all conditioned to view life and death in a binary, black-and-white manner. Even when faced with overwhelming scientific evidence and odds to the contrary, we still try to stick to this outdated concept, hence the creation of new terms, like *clinical death*, which are, in fact, meaningless. You don't find intensive care doctors saying our patients are clinically dead. We either say they are dead, or we say they are being resuscitated back to life while in a state of cardiac arrest. And for the sake of clarity and since we know what we are referring to, we very often simply tell each other such and such person died, but we got him or her back to life again.

As we saw, the need to try to fit with social convention was the major reason why Dr. Sestan, Yale University staff, and the NIH, which had funded the research, felt compelled to downplay their findings to ensure they would fit with the prevailing social beliefs about life and death. Every step of the way, they avoided talking about the ability to reverse death. Why? Because it is too difficult to announce a discovery that will literally, from one day to the next, completely challenge the most fundamental fabric of our society—namely, life and death. The ramifications were too great.

As demonstrated by the ethical challenges that arose with Dr. Sestan's two studies in dead pigs, due to the rapid medical and scientific advances of our era, we have no choice other than to integrate the study of what happens to the mind and consciousness into this new and evolving frontier of scientific enquiry that lies beyond death. After all, we are all conscious, thinking beings, and it is this consciousness, our selfhood, that makes us all unique. Since death has now been discovered to be a grey zone and people can be brought back to life, we can't ignore the important question of what happens to their minds and consciousnesses as they transition into this newly discovered zone. What can those who return from death tell us about what happens to us all when we die? These questions have often been ignored or considered out-of-scope by the scientific community. But are they? If we can bring the brain back online after being dead for hours and days in the future, how can we ignore the need to study what happens to the entity that makes a person truly a person as they transition into death during this time?

Dr. Sestan, the NIH, and the Yale University administrators were rightly concerned with the potential to restore conscious awareness to those pig brains, as it could potentially lead to immeasurable mental trauma and suffering in those animals. They also recognized that what they were doing would pave the way for similar studies in the near future in humans, too. However, what these individuals likely didn't consider is the possibility that consciousness and selfhood may not be annihilated as people enter into the grey zone. What if the experience of death is completely unlike anything that science has ever seriously considered? Perhaps the experience in the grey zone has the potential to be awe-inspiring and enlightening. No matter what the experience, it is something that has fascinated humans since our first ancestors realized that "one day I will die." This is perhaps why this, too, has been explored for millennia. But again, it was only in the twentieth century that science started to make real inroads with this subject.

Chapter 3

EXPLORING DEATH: PAST TO PRESENT

IN THE THOUSANDS OF YEARS BEFORE THE ARRIVAL OF MODERN medicine, people suffered and died from illnesses that would seem trivial today. Ancient doctors could offer them very little: absurd diagnoses, primitive remedies, and no pain medication even for something as excruciating as an amputation or childbirth. They couldn't offer much in the way of hope or help for their patients either; however, they weren't the only group attempting to understand death, and specifically what happens to us at the moment of death. For centuries artists have tried to interpret what happens after we die. Many have tackled their own versions of hell, heaven, and paradise—from Sandro Botticelli's *The Map of Hell* to Michelangelo's *The Last Judgment* to many more. But few have tackled the process of dying or the journey itself. However, the famed Dutch painter Hieronymus Bosch, who painted between the years 1470 and 1515, did just such a thing in Visions of the Hereafter.

I am fascinated by one painting in this series, *Ascent of the Blessed*. Unlike other portrayals of the hereafter that show people either burning in hell or enjoying some version of paradise or heaven or worshiping God, Bosch's *Ascent of the Blessed* depicts a large tunnel, often described as a funnel, as it narrows and erupts into a blinding white light on the top of the otherwise darkly rendered panel. In the middle of the riot of light, a solitary, faceless figure awaits as a winged luminous entity, dressed in a white robe, guides a naked person through the darkness and grey clouds up toward the light.

Beneath the tunnel, there seems to be a waiting room of sorts filled with other human beings, each of which is being gently guided, suggesting that as they are pulled away from Earth, they need only one guiding entity as they become lighter—less attached to the body—and make their ascent. The luminous entities aren't simply surrounding or carrying the humans, they are seemingly giving them assurance and helping them along. Meanwhile, the people ascending seem transfixed and drawn to the white light that illuminates the darkness.

Over six hundred years ago, long before there was such a thing as cardiopulmonary resuscitation or modern medicine, Bosch seemed to have captured something of what people have now come to describe as a recalled experience of death—specifically, the experience of ascending toward a bright light, as if in a tunnel, while being guided by a compassionate being (interpreted and drawn by Bosch, a Catholic, as angels).

One must wonder whether Bosch had survived a life-threatening encounter with death himself, or whether someone else had described one to him? Unlike that of other artists of his era, his work does not portray contemporary religious beliefs or traditions about death. Instead, this painting by Bosch is different—it portrays an experience that is stripped of religious trappings and infused with some of the principal elements of a recalled experience of death: loving entities, a sense of journeying, an ineffable light, and a separation from your life on Earth as you knew it. As much as I like this painting, nothing about it particularly surprised me the first time I saw it. It was first shown to me by a patient who had been successfully resuscitated. He was struggling to describe the experience he claimed to have had while his heart had stopped. Eventually, he gave up and simply relied on this image.

Bosch was painting something I'd heard described many times, both as a doctor and a researcher focused on understanding the experience of people who've died and been resuscitated. Over the years I've met countless survivors who've described a recalled experience of death. They are of all ages, beliefs, and backgrounds, yet there is a uniformity to their experiences. They all describe some variation of the scene depicted in this painting. One man shared, "I was being shown that the central sun was where all energy comes from and returns to, no matter where one calls home from lifetime to lifetime." They even describe figures very much like the ones Bosch painted. "I then heard that familiar voice, the one I had always heard behind me say the words... 'Hurry, we must go!' I immediately turned around... I saw the individual... can only be described as what must be an angel." And survivor after survivor describes the sensation of traveling under the guidance of an entity. These individuals are part of a long continuum of people compelled to talk about and share what they've experienced.

One of the oldest written references to unusual, recalled experiences surrounding death is from Plato's *Republic*, written in the fourth century BC. Here, a soldier, Er, suffers a near-fatal injury on the battlefield, is revived in the funeral parlor, and describes a journey from darkness to light accompanied by guides, a moment of judgment, feelings of peace and joy, and visions of extraordinary beauty and happiness. Plato of course contributed enormously to what scientists would later refer to as the problem of consciousness, *what makes us who we are*, which he and the ancient Greek philosophers termed the *psyche*. It is interesting to wonder how much this could have influenced his later thinking about what happens after death.

In the late nineteenth century, Sir Francis Beaufort, a highly respected and esteemed admiral with the British navy, wrote about his own lucid experience when he had almost drowned to death as a young sailor in Portsmouth harbor in 1795. He described this in the following manner: "Though the senses were... deadened, not so the mind; its activity seemed to be invigorated in a ratio which defies all description, for thought rose above thought in rapid succession."

He then described how he had relived every minute aspect of his life. He wrote, "A thousand other circumstances minutely associated with home were the first reflections. Then they took a wider range, our last cruise... a former voyage and shipwreck, my school and boyish pursuits and adventures. *Thus, traveling backwards, every past incident of my life seemed to glance across my recollection in retrograde succession.*"

In his article, he explicitly highlighted the immensity and depth of his own lucid review of life. He explained that "[those life incidents were experienced] not in mere outline," but "the picture filled up with every minute and collateral feature . . . the whole period of my existence seemed to be placed before me in a kind of panoramic review, and *each part of it seemed to be accompanied by a consciousness of right or wrong, or by some reflection on its cause or consequences*; indeed, many trifling events which had been forgotten then crowded into my imagination, and with the character of recent familiarity."

His testimony, written almost two hundred years ago, when virtually nobody had heard of these experiences, is consistent with what millions of people would come to describe in modern times.

The first systematic study of what people experience in relation to death was carried out in 1892 by Dr. Albert Heim, a Swiss professor of geology at the University of Zurich and director of the Geological Survey of Switzerland. Heim's work, especially his *Mechanismus der Gebirgsbildung* (1878), was highly regarded. He was awarded the Wollaston medal in 1904 by the Geological Society of London.

Heim, though, while an esteemed geologist, had survived a near-fatal mountaineering accident and had undergone an experience of death while unconscious during his life-threatening accident. He had also watched other people fall many times. He reasoned with himself that although it is terrifying to watch people fall, his own inner experience of coming close to death was different. This led him to systematically study other people's experiences. He wanted to offer some relief and consolation to the families of mountain climbing accident victims, by showing them that although falling to death is terrifying for those who observe the aftermath, from the perspective of the person who dies, the experience is likely not so.

After collecting and analyzing thirty testimonies, he presented his findings to the Swiss Alpine Club. These were later published under the title "*Notizen uber den Tod durch Absturz*" ("Remarks on Fatal Falls") in the

Yearbook of the Swiss Alpine Club. He concluded, "The subjective perceptions [of a person who faces death] . . . are the same whether [they] fall from the scaffolding of a house or the face of a cliff, . . . [or if a person is] run over by a wagon or crushed by a machine, even the drowning person, or he who senses himself falling on the battlefield, looks death in the face with similar feelings." He added, "There occurred, independent of the degree of their education, thoroughly similar phenomena, experienced with only slight differences. In practically all individuals . . . a similar mental state developed." Just as Admiral Beaufort had experienced, Heim also found evidence that the dying person experiences a review of their life.

Intrigued by Heim's work, Victor Egger, a French philosopher, put forth the phrase *L'experience de mort imminente*, or "imminent death experiences," in an essay, "*Le Moi des mourants*" ("The Self in the Dying"), in 1896. That same year, Egger's work was formally examined by the publication *Psychological Review*. They wrote, "Many persons who have survived an accident that seemed to be fatal report that at the time their whole past life came up before them." They concluded, "The civilized adult about to die and capable of reflection normally realizes his personality in a form vivid and significant."

Although Egger and others had recognized the need for a systematic study as far back as 1896, not much happened over the next eighty years. In fact, the general public and the wider scientific community remained almost totally clueless about people's recalled experiences surrounding death throughout the nineteenth and most of the twentieth centuries. This was in part because virtually anyone who would become so severely ill or injured as to lose consciousness on the brink of death would normally not live, let alone be able to recount their experiences. Of course, even if they had survived to talk about their experiences, most people in both the scientific and the religious communities would have dismissed them because what they had to say didn't match with their own notions and beliefs around death.

Additionally, in the nineteenth century, the reach of the media was also not as widespread or tuned in to these experiences. So even if some people had somehow survived a close encounter with death, their testimonies would have been recounted only to close family, and then would have been lost.

In the twentieth century, the world started to change drastically. By the 1950s and '60s, more and more hospitals in cities across the world became equipped with modern ventilators and life-support systems. Now people who had been rendered unconscious, while teetering on the cusp of death, could be kept alive because those life-support systems prevented critically ill people from deteriorating to a point where their hearts would stop and they would die. The game-changing field of intensive care (critical care) medicine and cardiopulmonary resuscitation were established realities. More and more doctors, nurses, and ambulance workers, even the public, had learned to perform CPR. Of course, most critically ill people still died, *but a growing and sizeable number did not.*

As mentioned, the key was that, for the first time in history, not only could people on the brink of death be prevented from dying, but in some cases even after biologically crossing beyond the threshold of death they could be dragged back to life again. As a result, many millions of people across the world who had been left unconscious due to the severity of their life-threatening illnesses or injuries were kept alive. *Although most did not retain any memories, around one in five to one in ten did.*

So, by the 1970s, it was just a matter of time before the almost totally unknown subject of consciousness and people's recalled experiences surrounding death would become globally known.

Somebody was needed to expand beyond the rudimentary work of Heim and Egger in the late nineteenth century. That person ended up being Raymond Moody, a medical student in his thirties from Georgia in the United States. With a PhD in philosophy and having taught the subject as a university professor before entering medical school, Moody was uniquely qualified for the task. He started probing his patients who had survived encounters with death about their recollections. He found 150 survivors of life-threatening events, including cardiac arrest, with profound experiences and set about studying fifty of them in detail. Noticeably, while those people did not know each other, their recalled experiences comprised a consistent set of what Moody identified as eleven key themes. They were:

1. *A sense of ineffability.* People consistently stated that they could not find adequate words to describe what they had experienced. It had felt like they had entered a different and higher dimension. As a result, ordinary language and words were insufficient to convey what they wanted to express.
2. *Hearing the news.* People described hearing themselves being pronounced dead.
3. *Feelings of peace.* A sensation that all was okay.
4. *Hearing unusual noises.* Though they couldn't explain them, they were not alarming.
5. *Going through a tunnel.* People reported moving through such a structure toward a light.
6. *Meeting others.* Some reported seeing loved ones or friends who had died in the past.
7. *Encountering a being of light.* They recalled a highly luminous, powerful, yet compassionate and benevolent entity who helped and guided them through what was perhaps the most intriguing aspect of their experience: the life review.
8. *A life review.* This took the form of a reappraisal and self-judgment of their own actions in life in a meaningful and purposeful way, viewed and analyzed from the perspective of ethics and morality. Specifically, it focused on how the person had treated others and what they had come to learn about themselves and their real worth as a human being—in essence, it stood as a gauge of their real humanity.
9. *Sensing they had reached a border or limit.* They knew that if they had crossed beyond that point, they could never have returned.
10. *Out-of-body experience.* Another intriguing aspect of the experience was that people described a sense of external visual and auditory awareness, seeing and hearing actual ongoing real-life events, during

their experience. This involved a perception of separating from the body while seemingly unconscious, yet observing attempts at being revived back to life by doctors and nurses. George Nuget Merle Tyrell, a British mathematician, had originally coined the term *out of body experience* in 1943. Moody used it to refer to this feature of the experience.

11. *Returning to the body.* Often, people recounted that they had been told they needed to go back to their everyday life, in order to complete unfinished work. However, almost always they did not want to return, because what they had experienced was so much more marvelous and powerful than anything they had ever experienced before. It was beyond words and, as Moody stated, was ineffable.

After recovering from their life-threatening event, many people felt their experience had provided a very profound and long-lasting positive impact on their lives. They became less materialistic, meaning they sought greater meaning and purpose in life beyond material and social measures of success or gain—such as wealth, status, or power. They became less self-centered and more altruistic. Yet when they had tried to tell others, they found that most people would not understand or believe them. Moody summarized these aftereffects into the following four themes:

1. *Telling others of their experience.*
2. *Effects on the person's life.* These were positive long-term changes.
3. *New views on death.* People became less fearful of death and came to view death as a transition into something new.
4. *Corroboration by others.* In some cases, other people had been able to independently corroborate and confirm the details provided by those who had described being able to watch real events from above while unconscious and on the brink of death.

Armed with five times as many survivors as Heim and with testimonies from people who had returned from much more profound life-threatening situations, Moody was able to identify many more themes than Heim. He noticed that these recollections bore a resemblance to certain philosophical notions of what happens after death. Since he had studied Greek philosophy before going to medical school and was familiar with Plato's story of Er, he was not at all surprised to find these same themes described by his patients.

In 1975, he published those testimonies in a book called *Life After Life*, which became an instant best-seller and garnered enormous global media coverage. Perhaps he had been unaware of Victor Egger's phrase "imminent death experiences," or perhaps he thought something different was needed. Either way, he coined a new phrase, *near-death experiences*, or NDE for short.

For the people who had undergone these experiences and many who learned of them, the specificity of the recalled features—consciousness with a sense of separation of selfhood from the body, combined with a meaningful life review—supported the age-old philosophical argument for the continuation of consciousness after death and a so-called afterlife.

Years later, in my own study, I would hear something similar from our own subjects. One, a British woman who had been bleeding profusely after a gynecological emergency, described how she suddenly found herself "high on the ceiling of the ward looking down upon the bed [which seemed to be a long way down] and saw the doctors and nurses around the bed working on the person lying there [which she then realized was her own body]." As her case illustrates, sometimes people don't initially recognize that they are observing their own body in this state. This is exemplified by someone who said, "I was floating above a person and didn't make the cognitive connection at first, that this was my body."

These testimonies seemed fantastical to some. Moody's contemporaries included scientists who claimed these experiences were either fabrications or hallucinations, delusions, or illusions arising from a dysfunctional brain in severely ill people who were at risk of death. They further argued that since death represents an absolute end, the human experience of death is not amenable to scientific study. Consequently, any recalled experience must reflect what happens "close to death" in a disordered brain, rather than an actual vision of what happens in relation to death itself. These debates continued onward for five decades.

I discussed the history of this subject with Dr. Bruce Greyson, a professor of psychiatry and neurobehavioral sciences at the University of Virginia, who has been studying recalled experiences surrounding death for more than forty years. He is the author of *After: A Doctor Explores What Near-Death Experiences Reveal About Life and Beyond.*

He explained that in the late 1970s many researchers and clinicians contacted Moody after all the publicity and interest from his book. So he decided to gather them together at the University of Virginia. The people who contacted him represented many disciplines. There were cardiologists, psychiatrists, psychologists, medical sociologists, philosophers, and scholars of religion. According to Greyson, "They all had different perspectives on what this phenomenon was and different favorite terms for what they were going to call them. One person argued for circumstantial logic experiences, another argued for Parthian experiences and so forth." He pointed out that some researchers coined alternative labels. "F. W. H. Myers called it a 'transitional dream.' Robert Crookall called it 'pseudo-death.' Jess Weiss named it a 'vestibule experience.' Douglas Hobson called it a 'perithan experience.' Bill and Judy Guggenheim coined the 'perimortem experience.' Craig Lundahl's was a mouthful: 'circumthanatologic experience' and so was Stephen Tien's 'thanatoperience.' Robert Smith came close to its actual meaning with the 'death survival experience,' as did Mike Sabom's 'actual-death experience.'"

Today, the French continue to use Egger's phrase, *l'experience de mort imminente*, and although other terms were also suggested later, including "spiritually transformative experience," nonetheless, as Dr. Greyson explained, "because Moody's book had already become so popular in the media, the term near-death experience is what stuck in the minds for better or worse."

When Moody and others had first gathered together, the scientific exploration of death, *rather than the period leading to death,* had seemed inconceivable. Based on social conventions, death was still firmly believed to be an unassailable and untraversable end. Dr. Bruce Greyson, who had joined that original group, explained that this belief was one major reason why people decided to adopt the term *near-death* experience. Their argument was that people can survive being "near to" death but not "actual" death.

Those early researchers, Bruce Greyson included, felt the final term would be of less importance because what really mattered were the details of people's testimonies and the circumstances around them. However, they failed to consider one inevitable consequence of not establishing stringent criteria at the outset regarding what it means to be "near to death." Bruce explained that Moody had specified that being near to death "has to occur during a close brush with death. [But] he didn't define [exactly] what that meant." In fact, in his book, Moody considered near-death to be any situation in which a person would have likely died without medical interventions. However, this was vague and not based on any specific objective biological determinant of being close to death. Yet clear biological and medical criteria of what it means to be "near to death" would be essential to avoid future subjective and erroneous misinterpretations.

This lack of foresight, while unfortunate, was understandable. Moody had been a philosopher and a medical student, and none of the other early researchers were specialists in life-threatening, or so-called near-death, situations either. To make things worse, some people then decided to arbitrarily remove even the improperly defined yet necessary requirement that any experience related to death should have a relationship with death, whether "imminent" as Egger had required, or "near" as Moody had stated. This is why, over time, a whole host of differing human experiences, which had little or nothing to do with being "near to death," or anything in common with Moody's original themes, or one another, started to be mislabeled as "near-death" experiences.

Common sense would dictate that human experiences unrelated to death are not "near-death." However, an unstructured "anything goes" approach, *the complete antithesis of scientific rigor*, had taken hold. This allowed more and more scientists and nonscientists, each with their own personal views, perceptions, and beliefs around what happens when we die, to impose their own beliefs rather than scientific realities on this subject. Science had truly turned into the Wild West.

Today, the medical literature is full of publications that purport to be about so-called near-death experiences. Yet most of them reflect experiences that have arbitrarily and subjectively been labeled that way based on the personal views of their authors, without proper scientific rigor. This is like a storekeeper who wants to believe apples and oranges are the same, and then satisfies him/herself by simply labeling them the same. These include, but are not limited to, experiences that occur in relation to fainting, sleeping, dreams, meditation, and the use of hallucinogenic drugs, such as N,N-dimethyltryptamine (DMT) and ketamine.

Nonetheless, from a medical science perspective, being "near to death" is a precise and objective event. You are biologically near to death when any life-threatening illness or accident causes your blood pressure to drop to dangerously low levels (also referred to as shock) together with loss of consciousness. This is when intensive care doctors are alarmed and working frantically to prevent you from actually reaching death. In medical shock, when your low blood pressure drops past a critical threshold, blood flow to the brain drops to dangerously low levels. This leads to confusion and, if severe enough, total loss of consciousness, meaning a deep coma.

Those observing a dying person from the outside often assume the absence of visible signs of consciousness and unresponsiveness also mirrors what is happening internally, meaning total nonexistence of consciousness. However, evidence was beginning to emerge that, during this period of a diminished and deepening state of unconsciousness, the person going through death may paradoxically be undergoing an *internal* transcendent, hyperconscious, lucid experience, which they would later either totally forget or partially forget and only recall in fragments.

Traversing from life to death can be analogized to entering an ocean, where life is land and death is the ocean. You start by dipping your toes into the water and getting a little wet while still standing on the land. This is the state of medical shock. As you walk deeper and deeper, you become submerged further and further. This is when the heart stops. Biologically and medically speaking, cardiac arrest and death represent an extremely severe (in fact, the most severe) form of medical shock. In other words, inadequate delivery of oxygen and nutrients to the brain and body, which contributes to cells becoming damaged over time. Biologically, death is an extension of what had started as shock. Eventually, when you walk far enough, you are completely submerged.

Throughout, from initial entry to having your heart stop to finally becoming fully submerged, it is all the same process of being in the water, basically a worsening severity of medical shock. However, deciding when you are completely out of land and completely in water is arbitrary. Is it after you are ankle deep, knee deep, thigh deep, or shoulder deep? It is not feasible, because walking into the ocean occurs as a continuum and in the early stages, even though you are partially immersed in water, you can potentially go back to land. Nonetheless, your presence in the ocean will remain permanent and irreversible if nobody can help pull you out. However, you can still enter deeper into this proverbial ocean. This is the grey zone of death. At some point, you become so deeply submerged that medical science will never be able to pull you out. Not today, not tomorrow, and not in a thousand years. This is when your dead body is damaged and degraded beyond repair, hours or longer into the postmortem period.

Yet for as long as your body is not yet fully damaged and degraded, biologically speaking, there is still a hope of resuscitation. The fact that the individual is not resuscitated simply reflects the limitations of medical science at restoring function in your organs and body. This could be because the necessary medical skills are not available in a given community or hospital, or because the necessary scientific discoveries haven't been made yet but will come about in the future.

This distinction is important, because it will help humanity search for ways to reverse death and restore life for millions of people rather than give up, because of an assumption that they have reached the "end." To put this into perspective, think again of all the otherwise healthy people who are declared dead after accidents, sporting activities, heart attacks, being in the cold, and so on. By considering death not so much as an end but as the beginning of a new process—an entry into uncharted territory—doctors can now start to actively look for new and novel treatments to "halt" and then "reverse" death before it goes too far. This fact is also what allows scientists to routinely go to a mortuary and take pieces from dead people's brains and grow living cells and mini-brains (known as *brain organoids*) from these samples in their laboratories.

Those organs in a cadaver are not damaged beyond repair, yet those cadavers are dead. Terms such as *imminent death* and *near death* and even *clinical death* were coined in the past to try to comply with the more old-fashioned and binary way that society tries to exert a clear line of separation between life and death. However, this is not compatible with twenty-first-century biological and medical discoveries. It is now time for society to change how it views life and death based on these scientific discoveries rather than trying to force its own arbitrary and outdated concepts onto science. But this is just the beginning. The story of the grey zone of death becomes even more intriguing.

Chapter 4

INSIDE THE DYING BRAIN: BURSTS OF ACTIVITY

In May 2023, news headlines once again reverberated around the world. This time in response to the work of Dr. Jimo Borjigin, a neuroscience researcher from the University of Michigan, who had set out to address two related questions: Is it possible for the human brain to become activated with death? Moreover, is it possible for signs of consciousness, awareness, and lucid thought processes—as an indicator of what Moody had popularized and millions of people had recalled of their experience of death—to be identified in people during *and after* death?

For Borjigin, the quest to answer these questions had started a decade earlier, in 2013. While working in the laboratory at the Department of Molecular and Integrative Physiology and Neurology at the University of Michigan, she had implanted six electrodes into the brains of nine rats. Then she and her team had begun administering lethal injections to each one. With the rats hooked up to machines, she had begun collecting detailed measurements of their brain activity as the rats' hearts slowed and they finally took their last breaths. She waited and watched as both the hearts and brains of these rats flatlined and ceased functioning. Then, to Dr. Borjigin's amazement, thirty seconds after the rats died, without a heartbeat, their brains suddenly and paradoxically experienced a burst of electrical activity. They seemed to go into overdrive and, in her words, showed "all the hallmarks not only of consciousness but a kind of hyperconsciousness." In an interview with National Public Radio (NPR) in the United States, she said, "We found continued and heightened activity; measurable conscious activity is much, much higher after the heart stops—within the first 30 seconds" after death.

Her work with dying rats had led to a lot of news headlines, and now, ten years later, she was generating headlines all over the world again. This time because she had started to study the brains of dying humans connected to an electroencephalogram (EEG) brain monitoring system. These are sensors that attach to one's scalp to detect electrical activity in the form of brain waves. She found that just like in the rats, electrical waves in the human brain had slowed and stopped with death. Then, intriguingly and inexplicably, a sudden burst of activity had emerged somewhere between thirty seconds and two and a half minutes after the heart had stopped.

Why would the brain shut down with death and then mysteriously be activated again afterward? With these results, Dr. Borjigin thought she had stumbled upon a discovery that would explain away the mysterious phenomenon that had intrigued scientists and the lay public for five decades. She was referring to the recalled experiences surrounding death, which she called "powerful experiences" when people are on the brink of death. She added, "They often say they had an overwhelming feeling of peace and serenity. Frequently they describe being in a dark tunnel with a bright light at the end. Many report meeting long-lost loved ones." She continued, "Many of them think it's evidence they actually went to heaven and perhaps even spoke with God."

Like some other scientists before her, she thought her study would discredit the validity of the millions of testimonies provided by people who had survived and recalled what Moody had termed a near-death experience. She thought her study proved the brain was simply doing what it does in order to, as she put it, "save itself." She posited that the brain was going into "hyperalert" to survive while at the same time making sense of all the brain cells firing. She chalked it up to nothing more than an "intense version of dreaming," *in effect, an imaginary experience arising as a trick of the dying brain.*

Is that all these newly discovered and intriguing signs of "hyperconsciousness" with death represented? Just a random and strange dream? An imagined experience without reality that occurs with death? More importantly, why and how would the human mind go into such a state of hyperconsciousness with death? Surely, as the brain starts to degrade and die with death, there should be a sense of diminished and fading consciousness, marked by slowing of brain waves on EEG monitors, *not paradoxical signs of lucid hyperconsciousness, marked by high-frequency brain waves.* What would be the evolutionary benefit—*the purpose*—of such a seemingly bizarre occurrence? What had science really unraveled regarding the mystery of what happens to us all, *to our consciousness and selfhood*, when we die?

Sara Reardon, who had also covered Dr. Nenad Sestan's breakthrough, wrote in the medical journal *Science*, "Burst of brain activity during dying could explain life passing before your eyes: New study hints at how consciousness can continue after the heart stops." Clare Wilson, from the *New Scientist*, said, "Brain activity of dying people shows signs of near-death experiences." Hundreds of other similar headlines flooded across the world.

Although Dr. Borjigin believed this was "maybe the first study to really show second-by-second how the brain dies," the claim to be "the first" was not entirely correct. Other scientists had previously reported similar bizarre findings, too. In fact, the tentative and presumptive link between "hyperconsciousness" and "lucid powerful" recalled experiences surrounding death and surges of electrical activity after death—*as have now been reported by millions of people across the world*—had first been put forth almost fifteen years earlier.

In 2009, Dr. Lakhmir Chawla, an intensive care doctor then working at George Washington University Medical Center, and his colleagues examined brain waves in seven critically ill patients who had died; Chawla noted a mysterious and unexpected surge in electrical activity after death. He initially dismissed his own findings, as he couldn't make sense of them. Surely, "the brain cannot be activated after death," he had thought. "That goes against everything we have ever presumed about death." But after he studied the brain waves on the EEG machines in more detail, he surmised that it wasn't simply a seizure or a "last-gasp attempt" of the brain to save itself. The reason for the sudden change in his thinking? *Gamma waves.* As he looked at the EEG results, he realized the electrical signal was of a very high frequency. "To see gamma wave activity, which is associated with consciousness, was massively unexpected," Chawla later explained. After he published his findings, he received phone calls from doctors all over the country who reported the same phenomena. "I don't think there's anyone who can argue that this is not happening," he later said.

In 2016, Dr. Loretta Norton, a neuroscientist at the University of Western Ontario, also reported a similar "unexpected finding." She found that "bursts of activity persisted" in the brain of a man after he had died. In 2021, Dr. Jan Claassen from the Department of Neurology at Columbia University identified similar findings—this time extending up to five minutes after death. Then, in 2022, Dr. Ajmal Zemmer, a neurosurgeon at the University of Louisville, found that thirty seconds before and thirty seconds after the heart had stopped, bursts of gamma waves had appeared in the brain of an eighty-seven-year-old man who had died.

Intriguingly, the longest report of electrical activity returning in the mammalian brain after death came in 2018 from Italian scientists, Dr. Pierpaolo Pani and his colleagues, from Sapienza University in Rome. They had implanted electrodes deep inside the brain of a monkey who was being euthanized and saw some sort of electrical activity, nothing like the high-frequency beta or gamma waves that characterize high levels of consciousness, up to two hours after the heart had stopped and the animal had died. This finding was totally inexplicable and remains so to this day.

Although we are witnessing the opening of a new frontier of science beyond death, even now the significance of the discovery of the electrical signatures of "hyperconsciousness" and "lucid" thought processes during *and* after death in humans cannot be overstated. After Dr. Zemmer's study in 2022, *Vice* wrote: "This isn't just about who was first to what: A case study like this, while intriguing on its own, is even more interesting when it's considered in combination with evidence offered by past work that has shown the same unexpected surge in electrical activity in the dying brain."

Brain oscillations, or waves, are patterns of coordinated brain electrical activity; beta and gamma waves are associated with the highest levels of conscious states, including learning and memory. According to experts from *Scientific American*, some researchers speculate "that a fully functioning brain can generate as much as 10 watts of electrical power," and so with enough electrodes connected to the brain, "you might even be able to light a flashlight bulb." This electrical power occurs in "very specific ways that are characteristic of the human brain. Electrical activity emanating from the brain is displayed in the form of brainwaves."

There are five major categories of brain waves, ranging from the most activity (highest frequency) to the least activity. When the brain is most actively engaged in mental activities and consciousness, scientists can detect its signatures in the form of increasingly high-frequency brain electrical waves—these especially include beta and gamma waves, the two fastest-frequency brain waves. The frequency of beta waves ranges from 15 to 40 cycles a second, whereas gamma waves can be as high as 140 cycles per second. According to *Scientific American*, "A person in active conversation would be in beta. A debater would be in high beta. A person making a speech, or a teacher, or a talk show host would all be in beta when they are engaged in their work."

Importantly, very-high-frequency gamma waves are found during moments when people are experiencing even higher levels of consciousness than with beta waves, a sort of state of hyperconsciousness. This is why Drs. Borjigin, Zemmer, and Chawla had all tried to link the discovery of very-high-frequency brain electrical surges after death to the long-standing reports of "hyperconsciousness" and "lucid" thought processes of the many millions of people who recalled experiences of death after being brought back from death. Dr. Zemmer explained, "Something we may learn from this research is: although our loved ones have their eyes closed and are ready to leave us to rest, their brains may be replaying some of the nicest moments they experienced in their lives."

These scientists had failed to directly link the high-frequency brain waves at the time of death to the hyperconscious lucid experiences reported by survivors. This was because the people involved in their studies had not been brought back to life to recount their experiences, and so it was impossible to link the two together definitively.

Nonetheless, as Dr. George Mashour, a colleague of Dr. Borjigin at the University of Michigan, had stressed, the surge of electrical activity isn't just the brain going "haywire" before death either. In contrast to what some scientists had claimed before, this wasn't just a "seizure," or "misfiring," or an "awry" brain causing a random, haphazard experience with death. Why? Because what was being recorded was coordinated, and in a specific higher wave frequency: the gamma bandwidth.

To Dr. Mashour, this connected the work done in animals to what's been observed repeatedly in humans, because "these studies mutually complement and support one another." Putting it all together, Dr. Borjigin explained, "If you talk about the dying process, there is very little we know . . . These gamma waves could mean those patients who had died had experienced some form of consciousness, similar to being in a lucid state."

Although no one knows what those handful of people really experienced with death, the research by Drs. Borjigin, Zemmer, Norton, Claasen, Chawla, and their colleagues has nevertheless revealed something extremely significant and extraordinary about what happens when we die—the paradoxical emergence of brain signatures of lucid consciousness after the brain has flatlined with death. Now to see whether they were truly related to the mysterious and powerful recalled experience of death, which Dr. Borjigin and others were referring to, large-scale studies were needed in people who were being actively resuscitated back to life after their hearts had stopped.

A FEW MONTHS LATER, ON SEPTEMBER 14, 2023, AN AVALANCHE OF headlines once again crisscrossed the world. This time the focus had turned to AWAreness during REsuscitation (AWARE-II)—the largest and most comprehensive study of the human mind, brain, and consciousness during cardiopulmonary resuscitation (CPR)—published in the scientific journal *Resuscitation*. I was the lead scientist and director of this program, which had been run by the New York University Grossman School of Medicine.

Ten months earlier, we had been invited to present our findings to a group of international scientists and cardiac arrest experts—and were one of only a handful of resuscitation studies presented in the "Late-Breaking Clinical Trial and Resuscitation Science Studies" session, selected from among numerous global studies by the scientific committee of the prestigious annual American Heart Association Resuscitation Science conference.

AWARE-II was truly a first-of-its-kind study in terms of its sheer scale and what it accomplished. It represented a collaborative effort among twenty-five major medical centers through a multidisciplinary consortium of thirty-three leading scientists and medical specialists in intensive care medicine, emergency medicine, neurology, neurophysiology, and anesthesia. Whereas previous studies of brain monitoring at death (by Drs. Chawla, Borjigin, and others) had followed fewer than 10 patients, we followed 567.

Importantly, this was the first time that any scientists had managed to incorporate tests of consciousness, as well as brain monitoring systems to peer inside the brains of people who were being actively revived back to life from the grey zone of death, to search for hidden signs of consciousness in real time. We could now try to identify those electrical brain waves, posited to be the signatures of a recalled experience of death, *live*, as they were taking place.

The 567 men and women in the study had entered the grey zone of death while undergoing CPR after cardiac arrest, during hospital stays mainly in the United States and United Kingdom. As with other cardiac arrest studies, due to the severity of this life-threatening condition, together with the limitations of conventional CPR methods (which haven't changed much since the 1960s), less than 10 percent recovered sufficiently to be discharged from the hospital.

To accommodate this reality, the study had a built-in complementary component. Additional testimony from 126 community survivors of cardiac arrest from around the world with self-reported memories would also be examined to supplement and build on what the survivors of those 567 cardiac arrest events would describe. Together they would provide greater understanding of the universal themes related to the recalled experience of death and what its specific components are, and whether they are similar to dreams or imaginary experiences, as Dr. Borjigin had said, or whether they represent a unique and real experience that emerges with death as millions of survivors have claimed.

Among the subset of 567 patients, our team of researchers planned to attach two brain monitors to their heads while being resuscitated. One measured brain oxygen levels and the other brain electrical signals second by second. There was also a novel and complementary way to test for consciousness. A tablet computer was attached above the head of the bed while people were undergoing lifesaving resuscitation attempts. It displayed random visual images and played specific sounds conveyed to the patients through Bluetooth-enabled headphones. Any positive recall of seeing the image or hearing the sounds would indicate the presence of conscious awareness as part of a recalled experience of death in people transitioning from life to death. However, the tests were designed such that even if people had no explicit recall of memories, it would potentially still be possible to determine whether they had experienced hidden elements of consciousness. This is referred to as testing for implicit, or unconscious, learning. There were very good reasons to think people had some form of consciousness and awareness but had forgotten parts, or even all, of their conscious experiences later due to the combined effects of massive brain inflammation (which happens when the heart stops but is greatly accelerated when oxygen is put back into the brain, as part of reperfusion, or reoxygenation injury, after the heartbeat is restored) and sedative drugs while recovering in the intensive care unit.

Due to the technical complexities of attaching monitors in real time during a major life-threatening medical emergency, 365 of those 567 people received the tablet and headphone combination, while 85 received these together with both brain monitoring systems.

AWARE-II found that one in five survivors of cardiopulmonary resuscitation after cardiac arrest had features of a recalled experience of death. However, almost 40 percent also had vague perceptions of awareness without any explicit recall. This supported our hunch that more people have consciousness and inner experiences while teetering on the edge of death, but later lose the ability to recall them.

The survivors' unique lucid experiences were structured and not haphazard. Instead, they followed a very specific narrative arc, including: a perception of separation from the body, observing events without pain or distress, and undergoing a meaningful, purposeful evaluation of their life, including all their actions, intentions, and thoughts toward others, followed by going to a place that felt like home and then returning back into the body.

Unlike many of our predecessors who have published works on near-death experiences using subjective labeling, which largely reflects personal views, we systematically analyzed survivors' testimonies using a scientific technique called grounded theory. This is a qualitative research method that scientists routinely use to explore people's beliefs, experiences, attitudes, behaviors, and social phenomena in depth. It allows scientists to gain insights into the underlying motivations, opinions, and meanings that guide human behavior. Unlike quantitative research, which focuses on statistical analysis of numerical data, qualitative research allows scientific analysis of non-numerical data, such as words, images, and narratives. It is a specialized research method that allows scientists to develop explanations for observations based on data that has been systematically gathered and analyzed, rather than mere speculation.

Using this method, we identified around forty other unique and consistent universal themes for the recalled experience of death in addition to what Moody had previously identified. *We also found this experience to be different from imaginary experiences, such as hallucinations, delusions, illusions, and dreams, as well as CPR-induced consciousness.* The latter is a term used to refer to the reports of movement, found in roughly 0.9 percent of all CPR events, that likely signifies that the heartbeat has been successfully restarted in people, even though doctors are still giving CPR.

Additionally, around 40 percent of the study participants, on whom our team of researchers were able to complete those first-of-their-kind second-by-second brain monitoring tests, showed signs of brain waves consistent with consciousness and some lucid thought processes at some point. *Though their brains had flatlined, they exhibited sudden spikes in brain activity even up to one hour into the resuscitation.* These were the so-called delta, theta, alpha, and beta waves. This was significant because, as already mentioned, some of these brain patterns ordinarily occur in people who are having lucid conscious thought processes and performing higher mental functions, such as memory retrieval and information processing. Now their detection for the first time in people undergoing CPR supported and validated the testimonies of millions of survivors of encounters with death, who like Admiral Beaufort, had recalled experiencing a state of lucid hyperconsciousness.

As mentioned, for decades, some scientists had claimed that people's experiences of death must be imaginary hallucinations or dreams that occur either before or after recovery from their encounter with death. They had argued it would be impossible to have consciousness while in a coma when the heart had stopped. Here, AWARE-II had disproven this. It also supported Drs. Borjigin, Zemmer, Norton, Claasen, and Chawla's observations of brain waves consistent with hidden elements of consciousness and lucid thought processes in people with death. However, it went much further by also determining the meaning and significance of these experiences.

Rachel Nuwer's article in *Scientific American* was one of the first to highlight the importance of these discoveries. She explained that the identification of "a flurry of brain activity during life-saving CPR" was a biological "sign of a 'near-death experience,'" which *NewsNation* and the Australian 9News network clarified were signs of "lucid consciousness," and "not hallucinations." These represented just a fraction of all the news stories that emerged across major media outlets from the four corners of the globe.

When asked about it, Dr. Lakhmir Chawla, the intensive care unit physician who had first discovered spikes of brain electrical activity at the time of death, was quoted by *Scientific American* as saying the study "represents a

Herculean effort to understand as objectively as possible the nature of brain function as it may apply to consciousness and near-death experiences during cardiac arrest." He was further quoted as saying he believed these findings could impact medicine today but should also "inform our humanity."

The AWARE-II study changed what it means to be on the brink of death and in a coma. As you will soon read, in our accumulation of studies of thousands of firsthand testimonies from people who enter the proverbial ocean of death, they describe journeying into a state of hyperconscious, lucid thought processes that they experience without fear or distress. In our AWARE-II study we measured electrical signs of lucid and heightened brain activity, and these, together with similar stories of recalled death experiences, strongly suggest that the human sense of self and consciousness, much like other biological body functions, is not annihilated when people enter the ocean of death and traverse beyond the traditional threshold of death.

This was one of those moments, and one of those discoveries, when you have to sit back in your chair and take a breath. We've all lost people we've loved—many of us have had, or are having now, moments where our own mortality feels imminent and inescapable. How would you feel if you could be confident that your dying loved ones, having left the shore of life, now wading into the grey zone of death, were at peace, or maybe even undergoing a wonderous experience rather than fear? These results—and the suggestion that our brains have one last surprise in store for us, and that there may be an experience beyond that which we expect, and even more astonishingly, that this concept holds up to scientific rigor and inquiry—were quietly ground shaking.

Our team of AWARE-II scientists translated these results into more academic speech and left the door open for further rigorous studies in the future. They concluded, "Although systematic studies have not been able to absolutely prove the reality or meaning of patients' experiences and claims of awareness in relation to death, it has been impossible to disclaim them either. The recalled experience surrounding death now merits further genuine empirical investigation without prejudice."

All of this had seemed completely inconceivable when I was trying to start my research career in the mid- to late 1990s. It all appeared like quite a daunting, if not impossible, task then. The subject seemed unsuited to scientific research, and everyone, including me, viewed life and death as separate, or binary, entities based on traditional social notions. Nonetheless, I realized that cardiac arrest would be the closest model for researching what happens with death. This made it challenging because the people we needed to study were the most critically ill patients in hospitals and were heading extremely rapidly, within a matter of minutes, to their death. Then like now, few would come out of this with their lives restored.

Where would I find enough survivors to make any study meaningful? It wasn't like I could go to a clinic and recruit patients as doctors normally do. I needed to find people who were actively undergoing a major life-threatening medical emergency. This was a highly unpredictable and relatively infrequent event that could occur without notice at any time of the day or night. It also necessitated timely medical treatments in a very tense situation—sometimes with up to a dozen people cramped into a small room working in unison. Nothing would be allowed to interrupt their emergency efforts.

Another major challenge was that then, as now, most doctors and scientists wouldn't go near this subject. It was and still remains somewhat of a taboo subject. I had to think carefully about the possible effects on my career. I was a young doctor just starting out and I didn't want this to possibly end my career, before I had even started it properly.

I was burning with enthusiasm, but with no research experience, I had a lot to lose. I persisted, though, and eventually a glimmer of hope emerged after I met Dr. Peter Fenwick, a neuropsychiatrist and neurophysiologist at the University of London. He was the only medical expert on this subject in the United Kingdom at that time and still is after thirty years. This shows you how little interest there had been in this subject then and now. During one of our meetings, he came up with what seemed to be a brilliantly simple yet practical way to get started. He suggested we use hidden images in an attempt to determine whether the people who claim to have seen things from outside their body had really seen those things, or whether they had just imagined seeing them. He demonstrated his idea by picking up a large white envelope from his desk. He drew lines and shapes on only one side of the envelope before standing up and lifting the envelope above his head. The side with the lines and figures faced upward toward the ceiling, and the side that was blank faced downward.

"If you look from below, what do you see?" he asked.

"Just a white envelope," I said.

"Exactly, and if you could look from above, what would you see?"

"The image and lines," I said excitedly.

This idea seemed so simple yet was so brilliant. We would mount images near the ceiling such that they would only be visible by someone looking down from above, but not when looking from the ground level up. This was important because the only details of people's recalled experience of death that would be amenable to objective testing were those claims made by people of watching doctors and nurses resuscitating them. It was testable precisely because doctors and nurses could independently verify any claims. But now there would also be an independent test, too.

In 1997, I officially started my research while working at Southampton General Hospital in England. For the next five years, I worked on two broad fronts, more or less in parallel. First, with the help and support of Peter Fenwick and a prominent local physician, Dr. Derek Waller, we started a scientific research project to examine what happens to the human mind and consciousness, but also the underlying biological processes after the heart stops—when people are actively undergoing resuscitation. Second, I started to study the testimonies of survivors of life-threatening illnesses and accidents, including cardiac arrest, from the wider community.

During an appearance on a national television program, I invited people to write to me with their experiences. I didn't know what to expect, but I waited in anticipation. Within a couple of days, I received forty to fifty letters, and the letters continued coming in daily for almost two weeks afterward. Within about fourteen days, I had received around 250 testimonies, and by the end of the year there were over 500 cases sent to me. These letters were truly enlightening. I read them all. Up until then, my knowledge of people's recalled experiences surrounding death had, like everyone else's, been limited to what I had read in books or scientific publications or heard secondhand from others. These firsthand accounts provided a real eye-opener and took my understanding of the subject much further. Many of those who wrote accepted my invitation to attend an interview at the hospital. By 2002, I think around 1,000 testimonies had been sent to us. I managed this large volume of correspondence with the help of some amazing volunteers, in particular Heather Sloan—a survivor herself, with a remarkable recalled experience of death.

At the same time, most of my effort went to setting up the study to test the main theories proposed for the causation of recalled experiences surrounding death. No one had ever done so. I really wanted to test the dying brain theory by examining whether lack of oxygen, or increased carbon dioxide, or drugs given during hospital stay may show any relation with people's recalled experiences of death. I also realized that we needed to test the psychological theories by looking for any relationship between people's cultural and religious backgrounds and the features of their experience. Finally, as Peter had suggested, we needed to independently test people's claims of being able to "see" their own resuscitation—the so-called out-of-body experience.

This was a very small pilot study that ran for one year. With the help of Becky Yeates, a research nurse, we recruited sixty-three cardiac arrest survivors, and fifty-six of them—around 89 percent—had no recalled memories whatsoever of their period of unconsciousness. Nonetheless, we gained significant insights from the remaining seven survivors (11 percent) with memories. Six of them described features of a recalled experience of death. All of the survivors were hesitant to talk initially. One lady's response in particular typified the very real concerns that people had about what other people might think about their experiences. She was so afraid of her husband's reaction that she made us promise that nobody else would find out about it. Eventually, days later, she confided in us about her experience.

We didn't find any support for the dying brain and psychological theories. Studies in the ensuing decades would also fail to find support for these theories. In the final scientific article, published in 2001, we concluded that "a continuum of near-death experience phenomena" occur. Our findings suggested that people's recalled memories could not be assumed to represent the full experience. In the same way that we all may experience many things—conversations, journeys, momentous events, and so on—in life, but later recall only a very small fraction of them, often in fragments, in death, too, people seemed to recall a very limited amount of what they had experienced and were describing their recollection as fragmented memories. Some people recalled more features than others. It wasn't clear to me then why there was a problem with memory recall. It took decades to appreciate why memory loss would be expected.

There was one anomaly that stood out. One person had a memory that showed no features of a recalled experience of death. It sounded like an ordinary dream of sorts. He had seen some people jumping off a mountain. The assumption had been that all recollections after a cardiac arrest represent the same thing—a recalled experience of death, or so-called "near-death experience," as Moody had labeled it. So we didn't expect to find other experiences, too. I didn't know what to make of it. But I made a mental note. It would take more than twenty years and the results of AWARE-I and AWARE-II to finally understand what that man's experience had signified.

During the one-year period that the study was being conducted I had painstakingly spent my weekends and holidays mounting boards with images that were only visible from the ceiling, while constantly changing and cleaning them. Yet none of the patients in our pilot study experienced the out-of-body phenomenon. So none of the images on the boards were ever able to be used. We learned the experience of separation from the body—the so-called out-of-body phenomenon—was quite rare and a much larger study would be needed to research this properly and put Dr. Peter Fenwick's idea to the test.

It took more than ten years to fund, set up, and eventually complete that large-scale study. This was the original AWAreness during REsuscitation (AWARE-I) study, which ran from 2008 to 2012 across fifteen mostly British and American hospitals. Our plan was to use hidden images on a much wider scale and study the testimonies of many more cardiac arrest survivors to better understand them. To test Peter's idea, we installed images on the tops of metallic fireproof shelves, which were attached to the wall near the ceiling. Because of the size of this project compared with the initial pilot study, we needed to mount some 7,500 shelves, based on the fact that each hospital averaged five hundred beds. We had neither the funds nor the staff to install that many shelves. So we decided to compromise and be tactical by identifying where patients were more likely to have cardiac arrests—such as in the emergency room or the coronary care unit. We estimated that if we covered those critical areas at each hospital, we would capture at least 80 percent of all cardiac arrest events. In all, we installed about 1,000 shelves.

In AWARE-I, 2,060 cardiac arrest events were recorded across the participating hospitals. However, again the vast majority died, and only 101 patients survived and were able to be fully interviewed. Although none had shown any visible signs of consciousness, or being awake, during their cardiac arrest resuscitation, 38 percent had a vague perception of being aware without the ability to recall specific details. As with the first study, there were also some people who, like the man who had seen people jumping off a cliff, also reported unrelated haphazard memories. These were again different from the classical experience that Moody had talked about. It was initially perplexing but started to make sense. People were forming other types of memories later on while they were in the hospital that were not of a near-death experience. Typically, they would stay in a coma for days or weeks after their cardiac arrest, so why wouldn't they form other memories later, too?

We found some of their recollections were simply dreams (e.g., seeing animals or plants), while others were disoriented memories of medical events that had taken place while they had been in a twilight zone—somewhere between being deeply unconscious and being fully awake. That is how we began to understand that not all memories reported by people after waking up from a life-threatening situation were a recalled experience of death (or "near-death" experience, as Moody had termed it).

In fact, soon after Moody's work was popularized, some people had arbitrarily claimed that there were distressing, negative, or hellish near-death experiences, too. But the evidence from AWARE-I (and later AWARE-II as well) showed these claims represented nothing more than frightening memories formed by people afterward when they had been disorientated and not fully awake. For instance, if doctors or nurses had been holding them down, they had mistakenly thought they were being attacked by "frightening creatures." If they had undergone a painful medical procedure, they had mistakenly thought the pain was because they were "burning in hell." These were not at all like the experiences that Moody had identified. In the medical literature these experiences had been categorized as intensive care unit (ICU) delirium. But some people had arbitrarily decided to label these as frightening or hellish near-death experiences without any scientific rigor.

Importantly, in AWARE-I almost 9 percent had features of a recalled experience of death and 2 percent—two individuals—actually recalled hearing specific sounds or conversations and visually observing their own resuscitation efforts from above. They had experienced the sense of separation from the body—the elusive so-called out-of-body phenomenon. While one patient declined to return to the hospital for detailed interviews because of poor health, the other, a fifty-seven-year-old man, accurately described details of events that had occurred in the cardiac catheterization room after his heart had stopped following a major heart attack.

He specifically recalled feeling that he had been above his own body. He said he had seen people in the room around him and that they had given his heart electrical shock treatment (defibrillation) twice. He had a bird's-eye view of all that was happening to him below while looking from above. He also specifically heard a nurse ("Sarah") speaking, and he also heard the staff say, "We've got him back." Importantly, we were able to verify that his period of conscious awareness had happened when his heart was not beating and had lasted at least three to five minutes. This was very significant, as it was the first time that a study had identified a case of conscious awareness—in support of what so many people had claimed to have experienced—while there was no heartbeat. It had also been able to show how long consciousness had lasted. In short, this was definitive verification of the perceived out-of-body phenomenon that people had talked about.

Nonetheless, after installing 1,000 shelves and following 2,060 cardiac arrest cases over ten years—which had yielded just two out-of-body cases—with our luck, both of them had been in areas of the hospital without a shelf! So our research staff were unable to ask if they had "seen" any of the independent objective images; and once more, the images were not able to be used. This is the reality of very low rates of survival after cardiac arrest, combined with the rare recall of the out-of-body phenomenon among survivors.

However, our findings did support the results of another significant scientific study that had been published in 2001 in the *The Lancet*, a prestigious medical journal, by Dutch cardiologist Dr. Pim van Lommel. He and his team had studied 344 cardiac arrest subjects and found one patient who had also reported a so-called out-of-body experience. As the man's mouth was opened to insert a breathing tube during CPR, his doctors noticed that he had dentures. One nurse then removed them quickly and placed them in a specific drawer before continuing to help with the resuscitation. After ninety minutes the man's heartbeat was restored, and he later recovered.

A week later, he was transferred back to the ward where that same nurse happened to be working. The man recognized her, even though he had been unconscious the entire time during his CPR. This really baffled the nurse. He then recounted where his dentures had been placed. He later told Dr. van Lommel that during the cardiac arrest: "I was floating up near the ceiling, and I was trying to let everyone know I was still alive because I was afraid, they were going to stop trying to resuscitate me." Based on this description alone, he, too, had likely maintained conscious awareness for some minutes while his heart was not beating and he was undergoing CPR.

AS SCIENCE AND SCIENTIFIC METHODS AND TOOLS HAVE PROGRESSED over the course of my thirty-year career, so, too, has the complexity and depth of our research projects. Aside from the use of sophisticated state-of-the-art brain monitoring equipment that has allowed us to peer inside and examine what happens to the brain second by second while simultaneously testing for lucid hyperconsciousness in the AWARE-II study, we have pioneered other studies, too. In particular, we have used state-of-the-art natural language processing (NLP) and artificial intelligence (AI) methods, in collaboration with our colleagues at the NYU Center for Data Science.

Natural Language Processing is a subfield of artificial intelligence, which uses computer-based complex mathematical models to analyze and process human language in various forms, including written text and spoken language. It requires huge sets of human data such as testimonies to learn what people are trying to express. In studying recalled experiences of death, it is superior to other methods that researchers had used when trying to derive meaning from people's testimonies. That is because it uses objective mathematical methods, rather than subjectivity, to understand and interpret human language in a way that is both meaningful and contextually relevant. However, to work in an accurate manner, it requires many thousands of datasets—meaning human speech or testimonies. The use of NLP has now been integrated into everyday applications, such as our phones and computers, which is how they recognize what we mean when we talk to them or ask them questions. When these technologies were first incorporated into our phones and computers, they were less accurate; however, the more that people have used them, the more accurate they have become at understanding the context of what we are saying. This shows you how accurate NLP has now become at distinguishing between what people are trying to explain as well as the power of its predictive accuracy when large volumes of human data are given to it.

In order to further objectively test the claims that recalled experiences of death may be imaginary experiences, like dreams or drug-induced hallucinations, as opposed to real experiences, which is what millions of survivors have claimed, we used NLP with large datasets of 1,251 testimonies derived from people with recalled experiences of death, 1,315 testimonies of dreams, and 1,200 testimonies from people with drug-induced hallucinations. This is the largest study to mathematically examine these three categories of experiences, and it showed, with 98 percent accuracy, that the newly discovered recalled experience of death represents a real and unique lucid experience that is different from dreams and hallucinations or other imaginary experiences.

Overall, all of our research work has been carried out in collaboration with a multidisciplinary team of investigators, including intensive care physicians, neurologists, psychiatrists, neuroscientists, neurophysiologists, and data science experts across multiple other major universities. This body of work has enabled us to glean much more knowledge about the features that make up the human recalled experience surrounding death, and our research results have been presented at many prestigious international medical conferences, most notably at the European Resuscitation Council meeting in 2018, the American Clinical Neurophysiology Society in 2021, and the American Heart Association Resuscitation Science Symposium (AHA-ReSS) in 2022 and 2023—the premier resuscitation scientific conference in the world.

We have now analyzed much larger samples of testimonies—starting with hundreds and now thousands of survivors of recalled encounters surrounding death—using much more advanced and sophisticated scientific methods and tools than those available in the twentieth century. Furthermore, our scientific understanding of life and death and the medical science around intensive care medicine and resuscitation—the science of bringing people back to life—is far more advanced now than it was in the 1970s, when people's recollections had first come to light, and when intensive care medicine was still very much a primitive and nascent field.

IN THE NEXT CHAPTER WE WILL EXPLAIN HOW THE AWARE-II study finally unraveled the scientific mystery of why and how the recalled experience of death occurs. Then in the next section of the book, we will explore in detail the different components of the lucid hyperconscious recalled experience of death, first by recounting a death experience, then some of the most astonishing examples of the recalled experience of death in detail, then by breaking down its components and illustrating them using first person examples from some of our accumulated data. Our understanding of the experience continues to expand, but a good place to start is with the main stages of the experience (which Moody had also found), including a perception of (1) separation from the body. This involves a sense of suddenly finding oneself outside of the body while maintaining full awareness and seeing and hearing precise details of what is happening. Over time, many similar testimonies were offered by different people, and their claims were later corroborated by doctors and nurses. There was also (2) a review of life, (3) travel to a destination, (4) arriving at a beautiful place, (5) encountering a bright and warm loving light, (6) seeing a luminous, compassionate being or deceased relatives, and (7) a decision to return to the body. These seven main features provide an overall narrative arc of the experience, which starts with separation from the body and ends with a return back to the body.

The lucid hyperconscious recalled experience of death is like a large, majestic tree with a trunk and seven main boughs, from which multiple branches, each with leaves and flowers, arise. Our work has helped identify these branches, leaves, and flowers, most of which had not been formally recognized before. These have started to provide a far deeper understanding of what the recalled experience of death really entails.

After analyzing thousands more testimonies, our team has discovered that the core themes that Moody identified, and the initial narrative arc, are just the beginning. In the same way that the beauty and uniqueness of a majestic tree can only be truly appreciated when viewed as a whole, rather than appreciating a leaf, or a branch, or a trunk in isolation, so, too, do the fifty or so themes, some of which are being presented in this book for the first time to the public, bring greater understanding of the powerful, lucid hyperconsciousness that comprises the human recalled experience of death.

Chapter 5

MYSTERY NO MORE: DISINHIBITION NOT DEGENERATION

YOU DON'T KNOW WHEN IT WILL HAPPEN, ONLY THAT IT WILL. Still, the moment itself catches you by surprise. You have errands to do: First, a quick stop to pick up the dry-cleaning. Death is the last thing on your mind. It's January, no snow, but very cold. Your spouse is shouting details of a last-minute task at you. You slip on the ice as you pull the driver's door open. Your spouse is still trying to get your attention—they are irritated over something and frankly you really don't want to hear about it. But as you pull out of the driveway you think better of your indifference. You stick out your head, look back over your shoulder, and shout, "What??"

That's when the car hits you.

Your last conscious thought is that your spouse was screaming, and a vague question: *What were they so upset about?*

Or maybe a long, slow illness brings you to the end in a hospital bed, surrounded by your most loved ones. Or, visiting a foreign country, you go swimming on a beach whose warning signs you can't be bothered to try to decipher. Perhaps you chew your dinner carelessly and feel a chunk of food lodge itself in your throat. You are alone, panicking, and you thump uselessly on your chest for as long as you can. At some point in all these scenarios, your heart stops. You have now put your first foot in the grey zone of death.

What happens to you next depends on multiple factors: How quickly your loved ones can get you help. How fast and well the paramedics do their jobs. The quality and availability of the equipment in the hospital. Whether the doctors and nurses look at you and think, *They have a chance.* How crowded and chaotic the emergency room is. And—more than anything else—sheer good luck.

The doctors who rush to your aid may well still believe ideas that are rapidly becoming obsolete. They were likely taught that the brain goes through a process of rapid degeneration after about five or ten minutes of oxygen deprivation with death. However, the reality, as Dr. Sestan's work has highlighted, is that often a brain that meets this criterion—meaning has lost function—still maintains *the potential* to be revived back for quite some time after death. Furthermore, when tests are carried out to determine the absence of brain function, they focus on brain stem reflexes, those reflexive functions that keep us alive, such as breathing or initiating a gag reflex (preventing objects in our throat from entering our lungs). Doctors cannot test for minute, detailed brain cell changes when they declare someone dead.

Nonetheless, scientists, doctors, and the public alike continue to view death as rapid degeneration, meaning a quick and permanent progressive and *linear loss* of all function and capabilities across the entire brain. Essentially, it is assumed the brain descends steadily and almost in a straight line rapidly from 100 (alive) to zero (death) within a short interval. A doctor who believes this is less likely to push to attempt resuscitation on someone whose heart stopped forty-five minutes ago, let alone hours ago, even though, as we will see in chapter 18, some people who have been in cardiac arrest for six hours have successfully been resuscitated.

AWARE-II showed how the idea of a rapid and linear process of degeneration of the brain after death is simply not accurate. On the contrary, it showed that the brain remains quite robust in terms of its ability to resist oxygen deprivation for prolonged periods of time after death. That is why even though, as expected, the brain flatlined—meaning it stopped broadly functioning—after the heart stopped, normal electrical markers, some of which were suggestive of consciousness, emerged even up to an hour later. This meant the nonfunctioning, silent brain had not degenerated, but had instead gone into a hibernation-like state with death, and the potential to restore function remained.

This is why by putting aside our preconceived notions of rapid degeneration with death, we can come to understand what really happens in the brains of people who enter the metaphorical ocean of the grey zone of death, where they report undergoing a recalled experience of death.

For our hypothetical dying brain, this means that those individuals whose brains and bodies are still physically intact, and who haven't signed a DNR (do not resuscitate) order, are still wading in the grey zone. So long as someone attempts resuscitation, they have a chance of returning to life. The brain itself can sometimes successfully restore life to the body, as evidenced by a phenomenon called "autoresuscitation." *This is because your brain is optimized to derive the greatest meaning for you out of every circumstance you face in your life and now also in death.* So it coordinates activities in your body to help it resuscitate itself back to life soon after death without any medical interventions, such as CPR. Although this is rare, it has been well described by doctors in the medical literature. (So next time you hear a news story about someone who had been declared dead but came back to life, unlike most people, who will automatically assume the doctor had make a mistake, you may appreciate this could have been a case of autoresuscitation.)

To better understand what is happening in a newly deceased person's body, let's go back to you, in the hospital, or wherever you had headed to death. No matter how it happens, your body—assuming it is still intact—is now in a desperate battle for life. Immediately, triggered by chemical signals released by the body, it enters a state of severe medical shock with unconsciousness—heralding entry into the proverbial ocean of death. This illustrates how the brain automatically attempts to optimize and derive the greatest meaning for the person. It switches from what can be called "life mode" to "death mode." As this state of medical shock worsens and the person descends further into the ocean and the heart stops and the person dies, the total loss of blood pressure, down to zero, further activates tiny receptors, which in turn trigger the brain to release even higher—*mega*—doses of several potent and potentially lifesaving hormones into the bloodstream. The period immediately after death is when the highest levels of adrenaline (epinephrine) are released by the body: one thousand times more than the amounts normally found in the bloodstream. As well as adrenaline, the brain also spews out steroids and other potentially lifesaving hormones, including norepinephrine and vasopressin, which work synchronously to try to raise the blood pressure by tightening blood vessels, squeezing more blood toward the heart and brain. These potentially lifesaving responses start when the blood pressure drops with medical shock but become far more vigorous in death after the heart stops.

Some of these hormones also act directly on the heart by trying to stimulate it to beat again. At the same time, after detecting low oxygen levels, the brain stimulates the lungs to initiate last gasp breaths. These are called agonal breaths and are an automatic reflexive response, seen in people as they die, that can help draw more oxygen into the body. But importantly, they create a vacuum in the chest that sucks blood away from the arms, legs, and abdomen and directly toward the heart and brain, where it is needed more. This is like an army calling up its reserves in the time of war. So, as mentioned, instead of an absolute loss of activity in a linear manner from 100 to zero across the whole brain in death, we see there is a dynamic process. There is dysfunction and loss of activity across *much* of the brain. This in turn facilitates activity in other normally dormant parts of the brain, which are better adapted to deal with the new reality of death. *Even now, the brain strives to make meaning out of this situation and its efforts are focused on kick-starting the heart back to life through autoresuscitation.*

These processes represent something we are calling "disinhibition," or the release of certain natural braking systems in the brain and body with death. In life, these braking systems run in the background of our brains, almost like a computer program that streamlines workflow, allowing the processes that matter the most, including your conscious mind, to focus on the things that need your attention in the moment. In death, however, there is a switching of activities. The brain relegates things that were of relevance to day-to-day life—for instance, what we eat, our bowel function, our reproductive needs, our relationships, our career, and so on—to the background. Instead, it elevates to the fore activities that existed in a potential yet dormant state. In this new reality, these functions become of optimal relevance and utility for people as they head toward and enter the grey zone of death.

It's important to understand how the brain adapts to life and death in order to understand the recalled experience of death. During day-to-day life, our brain is adapted to facilitate all our important needs. Yet in a moment of crisis, our brain can put all this aside and instead focus itself on what matters the most given the circumstances. In death that means trying to restore life back to the body.

One very simple example of this is how our brain automatically switches between two different modes of thinking depending on the circumstances at different moments in our life. In *Thinking, Fast and Slow*, the Nobel Prize–winning psychologist Daniel Kahneman explains how our brain enables us to operate using two contrasting modes of thinking. One, which he refers to as System 2, is slow, deliberate, and logical. We engage with this under ordinary day-to-day circumstances. However, when faced with severe pressure, stress, or frightening situations, our brain switches into what Kahneman calls System 1, meaning fast, intuitive, and emotional thinking. This mode is optimized to help us instinctively and rapidly deal with any perceived frightening or life-threatening situation. We automatically make very quick and instinctive decisions to avert danger in circumstances that are not suited for slow, deliberate thinking.

You can't, and wouldn't want to, carry out purposeful deliberate thoughts regarding your career, relationships, or other important aspects of your life when faced by a mob of angry people. When the brain switches to the fight-or-flight mode, it goes from everyday life to metaphorically confronting that mob. It does so through an active process of inhibition, by activating natural braking systems. These brakes block pathways that usually process information through the frontal lobes of the brain, which normally deals with voluntary actions and thoughts, such as elaborate decision-making and planning. Instead, all information and decision-making is diverted through the limbic or emotional center of the brain so you can make fast emotional decisions, which will help you to instinctively either run away out of the emotion of fear, or fight through the emotion of rage.

This is just one example of how the brain adapts to help us deal with different circumstances in an optimal manner and make meaning out of situations in life. It highlights inherent capabilities that often lie dormant in the brain but can be activated under the right circumstances. This dynamic system of inhibition of some parts and activation of others by the brain is like a system of traffic lights that automatically diverts traffic—in this case information processing—through optimal pathways depending on the circumstances. These inherent functions and capabilities are like sleeper cells that lie dormant and out of sight yet can be activated when needed.

This dynamic system of natural brain inhibition and activation is very important in allowing people to function optimally. For instance, imagine for a moment if all the mass of activities in your brain suddenly rose up to the level of your consciousness. You would become aware of everything happening in the brain all at the same time. The reality is that you would not be able to cope with the vast information overload and would become completely nonfunctional and paralyzed in your decision-making. The process of inhibition allows you to focus on activities that are relevant for a given set of specific circumstances.

So, in this manner, the brain is optimized to perform what you need under a given set of circumstances yet maintains an enormous potential for a host of activities that may be necessary and important at certain times. However, the vast majority of those functions lie in a dormant state ready for use if and when the specific conditions and circumstances arise. We may never even know we possess such an arsenal of potential abilities until something facilitates their emergence.

In the *New Yorker* magazine, the neurologist Dr. Oliver Sacks highlighted such an inexplicable case. Dr. Tony Cicoria is an orthopedic surgeon, who, in 1994, at the age of forty-two, had been struck by lightning. He had seen "a flash of light coming" before being hit in the face. This caused his heart to stop. He described separating from his body and watching events from above. Luckily, a nurse was close by and started CPR before an ambulance arrived.

Tony was successfully resuscitated and eventually he recovered. However, over two or three days, he noticed something strange. He had developed a new and insatiable appetite for the piano. Sacks explained, "This was completely out of keeping with anything in his past." He started to buy recordings and found himself especially captivated by "a Vladimir Ashkenazy recording of Chopin favorites: the 'Military' Polonaise, the 'Winter Wind' Étude, the 'Black Keys' Étude, the A-flat Major Polonaise." He wanted to play them all, and his head was flooded with music that seemed to come from nowhere. Cicoria taught himself how to play the piano, and within three months of the lightning strike he was spending nearly all his spare time playing and composing music on the piano.

Cicoria is not the only person to have developed remarkable, even superhuman, abilities after suffering a level of damage and dysfunction to the brain. Another more common example is seen with mania. When the condition is still mild (hypomania), people often describe a sudden surge in creative ability. But then, as their episode progresses and becomes severe, they lose that ability.

The key is that a certain level of brain dysfunction in some domains seems to bizarrely optimize and facilitate the unlocking of other inherent yet dormant potential activities in other areas. This seems to give access to hidden aspects of abilities that are inherent in the human mind and consciousness that would otherwise be inaccessible. Of course, there is a balance, a so-called sweet spot, and when the dysfunction and damage become excessive, then those abilities are lost. However, at a certain level, dysfunction unlocks them and gives access to hidden abilities in a person's consciousness and the mind.

Although the general perception is that damage and dysfunction in the brain is absolute, the reality is far more complex. Remarkably, dysfunction in some brain domains can unlock new and otherwise inexplicable, even superhuman, abilities. It is seen in people with developmental disabilities, such as autism spectrum disorder, or those with brain injury, who suddenly exhibit extraordinary skills or talents that are well beyond what would be considered normal. They are in fact highly extreme. People sometimes call this savant syndrome.

One of the most remarkable cases is that of Daniel Tammet, a now forty-five-year-old autistic Englishman well-known for his exceptional abilities in mathematics, language learning, and memory. One of his best-known achievements in mathematics is his ability to recite pi to 22,514 decimal places, a feat he accomplished in 2004, at the age of twenty-five. He also has proficiency in learning and understanding new languages very quickly. He is thought to speak several languages, including English, French, Finnish, German, Spanish, Lithuanian, and Esperanto.

In fact, Tammet is just one of a few people in the world who can perform complex mathematical calculations in his head. He can also learn a new language in a week. These abilities are attributed to his autistic savant syndrome. Somehow the dysfunction that has affected parts of his brain and caused autism, in turn, facilitated and unlocked incredible mathematical and linguistic abilities, which are unheard of among people with "normal," non-dysfunctional brains.

What is it that enables people with savant syndrome, like Tammet, to develop such extreme mathematical abilities, which none of us, not even the top mathematicians in the world with seemingly normal brains, can emulate? How is it that people whose normal brain functions have become dysfunctional can paradoxically develop superhuman and extreme abilities? Why do people with hypomania develop incredible levels of creativity and then lose them when their condition worsens? Why do such things occur? The answer is that when some areas of the brain experience dysfunction or loss of function, it causes disinhibition of inherent hidden pathways. This leads to activation of those otherwise dormant brain pathways, which unlocks access to parts of our own hidden consciousness. It also suggests human consciousness may be potentially far vaster than what we can ordinarily perceive.

This process of inhibition and disinhibition is seen elsewhere, too. Dr. Chandan Sen, a world-renowned regenerative medicine expert and pioneer of novel wound care technologies from the University of Pittsburgh, is one scientist studying the way genes can be inhibited and disinhibited in the body. He identified a series of genes whose function is to repair damage to the fetus body while it is in the womb. He explained to me that these genes are quickly activated if a fetus is cut. They then help repair and heal the damaged areas, so much so that after the birth of the baby you cannot find any evidence of a cut. However, if you make a similar cut in adults, then it leaves a permanent sign of damage. The reason is that after birth, those repair genes are shut down and inhibited. They are not activated again after a baby is born. Yet remarkably, Sen and his team discovered that the only time those genes are activated again is after death. Why? Through the process of disinhibition: the body automatically activates these dormant repair pathways, presumably to help repair any damage and help restore life again.

As we are seeing, then, with death, too, there is a loss of overall function—yet this overall loss of function is not absolute and across every minute domain. Instead, it leads to disinhibition and the removal of some braking systems. This in turn seems to facilitate new extreme abilities, which are otherwise inaccessible throughout the rest of our ordinary, "normal" life.

Once the brain dysfunction progresses beyond the mild phase to more severe states, people lose these abilities. A manic person will lose their creativity if their mania goes too far. People with extreme brain damage and dysfunction cannot do what people (like Tammet) with savant syndrome do. This likely happens in death as well, and people's newfound access to their deeper consciousness and other dimensions of reality may become inaccessible beyond a certain point. Here, too, there seems to be a certain spot where these remarkable functions come to the fore, and when the brain injury process goes further, we no longer see those abilities. Clearly, people who survive to report recalled experiences of death have not suffered with extreme permanent forms of brain damage, and have returned to life; otherwise they couldn't talk to us afterward. Like Cicoria, they had milder forms of brain injury and dysfunction.

What we have seen so far is that contrary to how people perceive death, meaning as an absolute end, the brain is adapted to deal with this new circumstance when it approaches and then enters into the grey zone. This is when those basic lifesaving brain reflexes—the ability to mount a gag response, or maintaining normal bowel movements—are lost. Electrical brain monitoring in people who are rapidly approaching and then entering the grey zone of death shows a slowing of brain waves, which then completely flatten within seconds. Then, after some time, we see those sudden surges of very-high-frequency brain waves before they are lost again.

These observations suggest there seems to be an inflection point of brain dysfunction (as we saw with dysfunction in life, such as with autism or brain injury), which triggers disinhibition and activates certain functions that were lying dormant in a sort of "sleep mode" before being lost again. But before that loss takes place, during that sweet spot, the brain areas that deal with what usually takes center stage in our ordinary lives—to help our relationships, careers, reproduction, bowel function, and so on—are no longer significant. They are inhibited. Our brain is adapted and optimized to help us deal with the new reality of being in the grey zone of death. In this dysfunctional brain state, through disinhibition, the brain ensures that the things that really matter in this specific situation for people take center stage, whether that be autoresuscitation or reactivation of those dormant repair genes. Meanwhile, your brain is also adapted to deal with something else: the reality that you are wading deeper into the grey zone of death and you are about to walk through a door with no idea of what you will find on the other side.

One of the key things to come out of AWARE-II was that it solved the mystery of why and how the recalled experience of death occurs and established a scientific explanation to put the earlier discovery of spikes of brain activity in and around death by Drs. Borjigin, Chawla, and Zemmer into context. AWARE-II concluded that as we die and our brain is shutting down and undergoing dysfunction, many of its natural braking systems stop working. This leads to disinhibition and allows for the activation of other parts of the brain, which had been dormant. This means that you access extreme yet otherwise hidden capabilities that provide you with an awareness of new dimensions of reality. This includes your entire consciousness, which has risen up like an iceberg and suddenly feels so much vaster. All your memories, thoughts, intentions, and actions toward others, from early childhood to death, are experienced not as random flashbacks but as a deep purposeful evaluation of your entire life. In this state, what matters to you is your moral and ethical conduct in every circumstance in life. The spikes of brain activity after the brain has flatlined are the "signatures" of this powerful inner hyperconscious and hyperlucid experience.

As dysfunction worsens, the optimal inflection point is passed. This is like a closed valve that is opened temporarily, before closing again. As we shall see, this explains why these abilities are thought to become snuffled out again when those who survive return back to their ordinary life. This also explains (aside from the issue of brain inflammation and the effects of sedative drugs), in part, why on their return people cannot recall so much of what they had learned in the state of vast hyperconsciousness with death (as will be discussed in chapter 13).

Nonetheless, this is how an overall state of brain dysfunction in relation to death, paradoxically, seems to facilitate the emergence of a unique lucid state of hyperconsciousness, as reported by millions of people across the world—not as a hallucinatory or imaginary state but as a real experience that emerges with death.

While no one knows the evolutionary purpose of why the human brain is optimized to facilitate the appearance of such a unique experience, we do need to ask ourselves: Why does our brain—that optimally makes meaning out of every circumstance in life—now in death also make meaning by facilitating this experience? What state of reality is it optimally preparing us for, and what is this experience that we are all going to have in that state?

Chapter 6

THE LUCID DEATH

TO START TO UNDERSTAND WHAT THE HYPERCONSCIOUS EXPErience we have been talking so much about feels like, let's go back to the emergency room. You've had an accident, like our hypothetical example. Or perhaps you have been fighting a slow-burning form of cancer for years. You're frail and elderly and your heart is finally giving out. Or you're young and strong, yet you've had a sudden calamitous health event. No matter what the cause, you're lucky you were not alone: someone, a stranger or a loved one, has called the ambulance and rushed you to the local emergency room. You are on a gurney, and all you can see is the bob and weave of masked faces over you. They are poking and prodding you, shouting orders. An IV is jammed into your vein, it hurts! Why are they so rough? Do they not realize that you are still "you"? Underneath the cold hands tending to you, and the crisp air-conditioned air, you are shivering and shaking. You are more scared than you have ever been in your life. Out of the corner of your eye you see a nurse walk by, scrolling her phone. This is the most consequential moment of your life, yet it is business as usual for most.

The edges of your vision are getting diffuse, yet you are still dimly aware of frantic efforts to save you, perhaps the push of medications, or an oxygen mask on your face. Dimly you might still hear your spouse, crying and pleading with you to come back. You feel intense panic as your brain fights for your life. Your awareness of what is happening shrivels away. Then suddenly everything goes blank. Your heart stops and you have no pulse. Without blood flow to your body and brain, you are no longer conscious and show no signs of life. You respond neither to pain nor your name. People around you consider you dead.

Although you do not feel any of their actions, you do find yourself even more conscious than you have ever been and somehow above and looking down at the scene of the accident or hospital bed. You see your own body lying in the hospital with doctors, nurses, or paramedics frantically working around you. The experience is surreal and bizarre and somewhat confusing. All kinds of thoughts rush through your head. *What am I doing up here? Am I dead? The people down below seem to think I am, but I feel more alive now than ever before.* You try to tell them, but they clearly can't hear you. Yet you can hear and see everything they are doing—but now you are viewing events, not as you had normally done, but instead from a 360-degree angle, all over and all at once. At the same time, you realize that you are not seeing or hearing the way you did before. Instead, it's a strange kind of knowing, like the sounds and images are flowing over and around you and in you, rather than being interpreted by your eyes and ears.

Despite the chaos and ensuing panic below, you feel absolutely no panic whatsoever. You can see the doctors working on your wounded and mangled body, perhaps with blood flowing or organs exposed, yet you feel no pain. Instead, you feel an overwhelming sense of peace and tranquility come over you. You sense you have shed your body, which had been weighing you down like an old coat, but still feel connected to it, by what you perceive to be a metaphorical cord of sorts.

In this new and liberated state, your thoughts are sharper than usual, faster than usual, more lucid than usual, and you continue to feel much more conscious and alive than usual. Unable to clearly fathom what is happening, you find yourself able to move around the room, or vicinity of the accident, and even beyond. You do this just by thinking and willing it. You then find yourself traveling at a vast speed toward a destination, through what feels like a passage, or tunnel of sorts.

You are drawn almost magnetically toward a radiant and comforting light. But this light is unlike anything you have ever experienced before. In part, this is because the "light" has a personality and character and feels very much alive. This is not simply a source of illumination. It is instead a personified luminous entity that exudes kindness, compassion, understanding, love, and warmth, yet at the same time is immensely powerful and knowledgeable and has a stature far greater than yours.

In the company of this entity, you start to relive every moment of your life—*all your actions, but also every thought and intention behind those actions.* You experience these from your own perspective, as well as from the perspective of those affected by your actions, together with the true reality of the matter. In other words, not as you had believed that matter to have been during life. *But as it had really been.* You feel and experience what others had experienced during your interactions—whether pain or joy—and come to recognize and experience the downstream consequences of your actions on others' lives. These moments may bring you immense joy, pain, sadness, or remorse, as you see, feel, and experience exactly how your choices and interactions impacted those around you, whether humans or animals.

You find yourself watching what you might later describe to be a movie of your whole life. Unlike ordinary situations though—where you can't recall most of the details of your life—now in this new state nothing is left out. You find yourself evaluating, judging, and reliving every minute event based on whether or not you had applied the principles of morality and ethics. Nothing is hidden, neither from you nor from the luminous entity accompanying you, not even your secret actions, hidden thoughts, or unexpressed intentions. You also view every time that you justified your own mistakes or unethical actions to yourself by blaming others for what they had done but refusing to take responsibility for your own part in it. You even find yourself evaluating the occasions when your intentions behind seemingly good acts had been self-serving, rather than truly for the benefit of others. Everything is laid bare for you, and you recognize that throughout your life, *from the perspective of morality and genuine sense of humanity,* you had often not been the person you had believed or had convinced yourself you were.

This review brings with it a profound sense of embarrassment and humiliation at times, but also understanding, as well as empathy, as you fathom the "cause and effect" relations behind your interactions and other people's actions. The luminous being who has been accompanying you throughout is also aware of your evaluation and tries to comfort and help you to understand the meaning behind your actions and your life.

Throughout this experience, your consciousness feels much more immense and vaster than anything you have experienced before; you feel like you suddenly know so much more than before. In fact, in comparison with your ordinary day-to-day consciousness and knowledge, it feels like you know "everything." At the same time, you recognize that the luminous being accompanying you knows far more than you do, and there is a hierarchy of knowledge, beyond even that of this seemingly "perfect" being.

As you move on, you reach a place that feels like home, somewhere you belonged and now have gone back to, which is permeated by an atmosphere of benevolence. The reality of a hierarchy of understanding and knowledge in this domain becomes clearer to you. The luminous being who has been accompanying you has an enormously higher level of understanding compared with you. You realize this being has an immensely higher field of perception. This is why it has a higher magnitude than you and is able to understand more truths. Yet you recognize that there are even higher beings, with an even higher field of perception, and far beyond them all lies an ultimate originating source, which is well beyond what you can fathom. You meet some of your deceased relatives, who may also appear as luminous entities—a personified light—but with far less intensity and field of perception than the glorious luminous entity who had accompanied you. Or they may appear in the form that you had known them throughout your life. Those family members greet and welcome you and bring you comfort.

By now, you have totally forgotten about your body down below and have lost interest in most of the things that had preoccupied you before:

your job, relationships, maybe even your family. In fact, that life, compared with what you are experiencing now, feels like a mere shadow of reality.

You then face a decision to continue or return to your body. As much as you are drawn to stay in this magnetic and marvelous domain that feels immensely more real to you, with its overwhelming sense of peace, love, and kindness that permeates the atmosphere you are experiencing, you are told that you need to return to your ordinary life. You gain a deep sense of appreciation that there are things you need to complete in your life as there is a deeper purpose and that your work there has not yet been completed. You feel like you are being prepared and guided back into your body.

When you regain consciousness, you find yourself again surrounded by the doctors, nurses, or paramedics who had been trying to rescue you. You suddenly feel the pain and discomfort of what is happening to your body. Despite this, after you recover, you recognize that since that day, your life has never been the same. This experience transforms you profoundly. It gives you a renewed appreciation for life, a deeper sense of seeking meaning in life, and a drive to live with more compassion, love, and understanding for others. You appreciate that your life—and specifically your interactions with others—has actually been an educational opportunity with a higher purpose to develop yourself based on some universal moral and ethical principles. You also seek greater purpose in life, while trying to understand what it was that you had understood you needed to accomplish now that you have a second chance at life.

Perhaps you also try to share your experience and talk about what has happened to you. Your loved ones, relieved and overjoyed to have you back, are open. It gives them comfort to believe that this experience is your abiding memory of your time in the grey zone. They may not fully "believe" you, yet they are happy to hear you talk. This may not be the case with your medical professionals—either the doctors who treated you then, or the people you see for follow-up care. You will be told that your brain was oxygen-deprived and hallucinating. That it came up with these images to block out the trauma you actually experienced. Another doctor might lecture you that brain waves and all function ceases after death, therefore this is not possible. If you are religious, your priest, rabbi, or other religious cleric might bristle at the sacrilegious imagery you describe. Eventually, you will stop talking about it, and get on with your life. You feel alone about your experience.

THE FACT THAT THE RAMIFICATIONS OF THE TWENTY-FIRST-century discoveries in relation to death by scientists such as Dr. Nenad Sestan have not yet been fully understood by society or the scientific community, together with a dismissive approach to the study of consciousness with death, leading to tension between the vividness of people's recalled experience of death and the refusal of many medical professionals to acknowledge their experiences, are just some of the reasons that led my colleagues and me to establish clear parameters and definitions for rigorous research. We also wanted to avoid the subjective mislabeling of a series of unrelated experiences with no relation to death as "near-death" experiences. This would help limit the criticism of survivors and those who study them, by filtering out experiences that did not meet the criteria for having any relationship with death. To this end, we formed the first-ever multidisciplinary conference to establish research guidelines for the scientific community.

This was held on October 26, 2019, at New York University, followed by a symposium and panel discussion on November 19, 2019, at the New York Academy of Sciences. Experts from the social sciences and humanities, resuscitation science, emergency and critical care medicine, psychiatry, psychology, neurology, and neuroscience explored the current state of understanding regarding cardiac arrest resuscitation, as well as what happens when people die and the initial postmortem period. They represented some of the world's most respected academic institutions, including Harvard University, Baylor University, the University of California, Riverside, the University of Virginia, Virginia Commonwealth University, the Medical College of Wisconsin, the University of Southampton, the University of London, and New York University.

Together, we published "Guidelines and Standards for the Study of Death and Recalled Experiences of Death" in the *Annals of the New York Academy of Sciences*. This multidisciplinary work examined the accumulated scientific evidence to date, and presented the first-ever peer-reviewed consensus statement for the subject. It put forth an appropriate definition of, terminology for, and a research framework for the study of these experiences. Additionally, it identified important knowledge gaps that will help standardize current research and establish future work by physicians, scientists, philosophers, and social scientists engaged with understanding what happens to human consciousness in relation to death and will help scientists distinguish between these and other diverse experiences. Finally, the group recommended the use of the more appropriate, accurate, simple, and descriptively correct term *recalled experience of death*, or *RED*, since often people with these experiences had entered the grey zone of death, rather than just being close to it—instead of so-called near-death experiences—and urged objective scientific research into the human mind and consciousness around the time of death.

Many of the people we interviewed have recounted that while lying unconscious in and around death, they suddenly developed remarkable abilities like those listed above. They gained knowledge and insights that are otherwise inaccessible under ordinary circumstances. Their consciousness and awareness—their inner mental universe, the thing that makes each of us uniquely who we are—became enormously vast. People were suddenly and inexplicably able to process millions of pieces of information about their own life, all in an instant and in a nonlinear manner. Their thinking became infinitely clearer, sharper, and more lucid than usual. As already mentioned, the loss of some brain function in death paradoxically seems to have enabled them to access other hidden realities, including their own entire consciousness and specifically with a focus on how they had conducted themselves in life.

Drs. Borjigin, Chawla, and Zemmer (and now the AWARE-II investigators) have all seen signs of this lucid hyperconsciousness after the heart stops. All of these cases, including those who died without resuscitation in Borjigin, Zemmer, and Chawla's studies, and people who were resuscitated in the AWARE-II study, had brains that were flatlined, and they all showed a sudden transient surge of brain waves normally seen in states of consciousness.

We all believe these brain-based markers are the signatures of the recalled experience of death. In the overall sense, Dr. Borjigin was correct. People do consistently report "hyperconsciousness," a huge and vast expansion of their consciousness with lucidity, and their unique experiences do provide insights into what the experience of death will likely be for us all. In Dr. Borjigin's case, she thought the explanation was simply the brain releasing energy, maybe even some sort of seizure that causes an imaginary experience, as a strange dream or hallucination of sorts. Dr. Lakhmir Chawla, the critical care doctor who has gone on to publish numerous scientific articles on the topic, was surprised and excited. As he said, there was "no blood flow and no oxygen going to the brain." If it was merely a dream or a hallucination, the brain would require blood flow and oxygen.

The AWARE-II study team determined, however, that what was happening was in line with the testimonies of millions of survivors with a recalled experience of death. Unlike the proponents of the dying brain theory, our team of investigators found an explanation that reflects what happens in the dying brain. We do not dismiss the reality of what people have recalled. We agree that they experience a new dimension of reality in death. We also do not think the finding of brain electrical markers of hyperconsciousness means the brain is producing the experience. On the contrary, it means the mind and consciousness, tethered to the brain in death, are interacting with and modulating the brain. Of course, this discussion goes to the heart of the ancient mind-body problem, also referred to as "the problem of consciousness" by scientists and philosophers today.

Although we still do not know the evolutionary purpose of this phenomenon, it reveals something intriguing about the mystery that is human consciousness in life and in death.

For Dr. Chawla, a fascinating aspect of these data and this topic in general is what people think the spike of brain activity after death means, and how it is interpreted often depends on their starting beliefs. He explains that this is probably true of all information that we receive: our biases and belief structures influence our reactions and interpretations in some way. But this topic, in particular, seems to be an apt example of that. "Having been through this so many times, whatever your belief structure is, if you wrote it down on a piece of paper before I showed you this data, I could predict with a high degree of certainty what you'll think of it," Dr. Chawla said. "It's an interesting notion about this data, but also about data in general" (the reason for this will be explored in more detail in chapter 18).

For Dr. Loretta Norton, ultimately it makes sense for there to be fascination around this topic. "We all die, and we're all trying to understand what happens to a loved one, or what happens to one's body during the dying process."

Now, almost fifty years after Moody had first popularized people's recalled experiences surrounding death, much more has been discovered and understood about the hyperconscious lucid experience of death, whose brain electrical signatures, in *the form of gamma waves*, Drs. Borjigin, Zemmer, and Chawla thought they had discovered. We also have a much deeper understanding now of what this experience feels like from the perspective of the dying than what people understood in the 1970s and even through to the 2000s. Similarly, we understand much more deeply what will happen when our own time comes to cross the threshold of death. We even understand the mechanism by which the brain facilitates this experience in preparation for what comes next with death.

In the next section of the book, based on the results of AWARE-II and our other studies, I will build a more detailed picture of the fifty or so unique features, which, when examined together like a tree in full bloom, give a much fuller picture of the unique features of the recalled experience of death than previously appreciated.

Section Two

EXPERIENCING THE GREY ZONE OF DEATH

Chapter 7

THE COSMOS: A VAST EXPANSION

IN THE LAST CHAPTER WE EXPLORED THE TRANSITION THROUGH death and recalled experience of a hypothetical patient. However, in 2019 I met a man who has waded in and out of the grey zone of death more than any other individual I have met or heard of. The hundreds of testimonies (many of which you will read shortly) were profound, illuminating, and convincing. However, meeting a man like Dr. Robert "Bob" Montgomery, a world-renowned physician and transplant surgeon who works with me at New York University (NYU), dramatically changed everything. Not only is Dr. Montgomery a top-notch surgeon, but he is a man of outstanding reputation. His willingness to discuss his extraordinary experiences in the liminal space between life, death, and beyond can change how people see and understand this field of research. Bob and I met after Dr. Stuart Katz, his cardiologist, himself a renowned physician and colleague, insisted that we talk. He thought—correctly—that Bob's experiences would be relevant to my work. It turned out that "relevant" was an understatement.

Aside from being the director of the NYU Transplant Institute, Dr. Robert Montgomery is a professor and the chairman of the Department of Surgery at NYU. He also has a PhD (D Phil) in molecular immunology. Before moving to New York, he had served as the chief of transplant surgery and director of the Comprehensive Transplant Center at Johns Hopkins for over a decade. He was part of the team that had developed the laparoscopic procedure for live kidney donation, a procedure that has become the standard throughout the world.

Obviously, I knew that Dr. Montgomery's professional accomplishments were second to none. Little did I know what an incredible life story lay behind it all. I also didn't know how much I would come to admire him, not just professionally as an accomplished scientist and transplant surgeon, but importantly as a remarkable human being, who has faced hard ethical and moral challenges. When we first met in his office, I found him to be a tall, athletic, larger-than-life man who looked in his late fifties. He clearly loved the outdoors and had a strong sense of adventure. Pictures of him on snow-covered mountains and in forests attested to his love of the wilderness. There were also signs of his professional accomplishments on the walls and on his desk. I saw numerous medical certificates, interlaced with photographs of him receiving awards from prominent people, including from President George W. Bush, who had honored him at the White House in 2008, after his historic six-way kidney swap. After we sat down, I realized that despite all his remarkable accomplishments, Bob was one of the most humble and genuine people I had come across. Then when he started to share his life story, I became transfixed.

He explained that his father became a mechanical engineer after flying bombers in the South Pacific during World War II and was involved in the Gemini space program. He had a wonderful childhood, but when his father was just fifty, he started having some heart-related symptoms. After a visit to the doctor, he found out he had a heart problem called cardiomyopathy.

Cardiomyopathy affects the heart muscle. People become progressively ill over many years as the heart starts to fail. Then they can die suddenly at any moment. That is because their diseased heart muscle develops an electrical malfunction, short circuits, and stops beating. It goes into cardiac arrest. Without medical interventions this also means permanent death.

It affected Bob's father—and his family—tremendously. His father had multiple cardiac arrests, several that he was resuscitated from successfully.

Then the last one resulted in him being in a chronic vegetative state (after suffering brain damage) for about four months. Bob had vivid memories of watching as a fourteen-year-old boy his once strong and brilliant six-foot four-inch, 225-pound father just slowly withering into this person who was in the fetal position being fed through a tube. He remembers mournfully, "We would visit him every day. It was just a horrible thing to watch. Then, at fifteen years old, I remember when one night at two or three in the morning we got a phone call that he had passed."

While many boys of Bob's generation were inspired by the Space Race to go into engineering, Bob was so moved by the death of his father that he became determined to find the answers to why his father was taken from him. Bob was less interested in the new frontier of outer space, and more interested in the unexplored human frontier that lies down on Earth.

The death of his father propelled him to get a degree in medicine. Then, twelve years later, when he was twenty-seven years old and working as a newly qualified doctor—an intern—at Johns Hopkins University, Robert received another call in the middle of the night, a call that would change the trajectory of his own life. It was a call no brother ever wants to receive. Bob's sister-in-law was on the other end and broke the news that his thirty-five-year-old brother, Richard, had suddenly and unexpectedly died. He had been water skiing, and he just let go of the rope, and when they circled back around to pull him out of the water, he was dead, and they couldn't resuscitate him. Having lost two family members now, Bob thought it was more than suspicious. His immediate question was: *This couldn't possibly be a coincidence, right?* He had his brother's heart sent to Johns Hopkins Hospital. There the pathologist looked at it and found evidence of "dilated cardiomyopathy." That was when he knew he had a big problem. The dots were connected. It was clear that he, too, was at risk for developing the same fate as his father and brother because the mysterious heart disease his father had died from was genetic. In short order, he, too, was evaluated. Sitting across from me in his office, he shared, "I too had cardiomyopathy."

It wasn't long after, in 1989, that under the supervision of doctors, Bob got on a treadmill and started running. When he got to peak exercise, unexpectedly he developed what is called ventricular tachycardia: a very abnormal and dangerous heart rhythm (which puts people at risk of sudden cardiac arrest and death). "I almost passed out," Bob recalled. "I still had a strong heart, but it caused great concern for the people monitoring the exercise test. They panicked and were hitting me in the chest, trying to get the heart rhythm back to normal."

By chance at the time, there was a surgeon at Johns Hopkins named Levi Watkins who had implanted the first defibrillator (called an ICD) inside a human body. Nowadays, they are the size of a coin and can be implanted near the shoulder. However, back then, this was a highly experimental and difficult procedure, not to mention extremely painful. Bob was the first surgeon in the world to undergo an ICD implant. His doctors told him having one of these devices would not be compatible with the life of a surgeon and he needed to change his career.

At the time, his chief of surgery arranged for him to go to the University of Oxford, in England, and work in research instead. He told Bob, "Let's see what happens when you come back." Bob ended up staying for three years and he earned his PhD in molecular immunology before returning back home.

Bob was doggedly persistent. With the support of his chief of surgery and armed with what he said the British refer to as his "bloody-mindedness," he refused to give up on his dream. In time he managed to singlehandedly test and prove that his ICD device would not be an impediment to him working in the operating room, and so after many years, he eventually became a transplant surgeon.

Bob said, "I had fully embarked on my dream career. Then over the next twenty some years, I had a lot of normality punctuated by near-death experiences, when I would nearly die, but thankfully they never happened in the operating room, or during any kind of a patient encounter."

In those twenty years, he married, had a family, and was productive at work. He got on with life, knowing this was a fragile situation that, in his words, "could turn on its head in a moment." In a way, it made his decisions about how to spend his time and focus his efforts more deliberate and purposeful. Living from one heartbeat to the next, he knew he had to, as he said, "get it right and nail my life—and in that regard my disease was a blessing."

However, as time went on, his heart became less efficient and started to fail. Then, like his father, Bob had his first cardiac arrest, in 2010, at just fifty years old. He recognizes that if it were not for his ICD, the event would have been fatal, much like it was for his own father and brother. At the time, he was climbing in the Andes, in Argentina, in a snowstorm, at a very high altitude with his son. He started to get very winded. They were in a blizzard, and he lost sight of the others. Next, he felt the abnormal rhythm coming on and he braced for it.

"Then there was nothing. I was gone."

He flatlined again in 2017 in Argentina and recovered. Then again, that same year, during a Broadway show, he had another. This time the ICD didn't work. Thankfully for Bob, a doctor and nurse in the audience gave him CPR and used an external defibrillator. With the Broadway show having to be stopped, they attempted to revive him for forty-five minutes while everyone in the audience watched. It was, according to Bob, a "real showstopper."

After all these episodes, you may be wondering, as I did, *Why didn't he just get a heart transplant? Why risk another inevitable encounter with death?*

Even though Bob Montgomery was the director of the Transplant Institute at New York University and in dire need of a heart transplant himself, he still did not qualify for one. He was not, in his own words, "considered sick enough." Most people in his situation would have prioritized themselves to get a heart transplantation. His only real option to save himself would have been to receive a heart transplant. He had already been brought back from death three times. He knew it was just a matter of time before it would happen again, and his heart was getting worse. There was also the real danger that like his father—even if he did survive the next episode—he could suffer catastrophic brain damage and go into a vegetative state. Yet despite all that immense pressure and the real danger to his own life—*literally with each heartbeat*—he resisted the pressure to try to push himself above others on the national waitlist. He could have tried. After all, no one would have argued that he did not need a heart transplant. He faced up to the challenge that others face—a scarcity of organs for transplantation.

It took another nine months for Bob to finally qualify for a heart transplant. In September 2018, Bob was attending a medical conference in the small town of Matera in Italy. Suddenly, he suffered four cardiac arrests over a three-hour period. He experienced something sinister called a ventricular tachycardic storm. He was in a vicious cycle. He would have a cardiac arrest, get shocked by his ICD device, come back, then go back into cardiac arrest again twenty minutes later. He later recounted how one of the first people he met in the hospital was a priest in the emergency room, who actually gave him the last rites. That is how dire his situation had become.

Eventually, the Italian doctors managed to somehow stabilize his heart and put him on several lifesaving drips. It was during this time that Bob asked one of his physician friends to help him get back to New York. He knew this "ventricular tachycardic storm" now made him eligible for a heart transplant, and he needed to get the transplant before his next cardiac arrest, which would potentially be his last. He had no time to spare. Against medical advice, he left the hospital. His Italian doctors were in shock when he told them, "I want to leave." But they were very compassionate and wished him luck. They even gave his friend, who was accompanying him, preloaded syringes of emergency medications—adrenaline and amiodarone—to treat him in case he had another cardiac arrest on the commercial flight home.

Thankfully, he made it home safely, but as soon as he arrived, he immediately called the heart transplantation team at New York University, where he was (and still is) the director of the Transplant Institute. He had just received the license to start NYU's first heart transplantation program, which he had initiated just six months before. He had also assembled a group of luminary heart transplant surgeons, cardiologists, lung specialists, and transplant team members who, unbeknownst to him at the time, would come to save his own life just one year later.

There was one more twist to Bob's story, though. As a transplant surgeon, for years he had sought to expand the number of organs available for transplantation because, as he explained, "perfectly good organs were being routinely discarded because they had come from people with HIV and Hepatitis C infections." He had been trying to find a way to safely give hepatitis C–infected organs to people. In fact, he and his team were starting to test his experimental protocol for the first time in the world when his own time for a transplant came. So he told his team, "I only want an organ from someone who dies with hepatitis C." His reasoning? "You shouldn't ask your patients to do something you wouldn't do yourself."

But there was something more pressing on his mind, too. He had guilt about competing for organs with his own patients. He said, "It troubled me a lot because when you get an organ, somebody else doesn't and that person might die. It seemed like the right thing to do." He was right to be worried. That same year, in 2018, around 114,000 Americans needed an organ transplant, but only 36,500 received one; each day, an average of twenty people died while waiting. About 4,000 people needed a heart, but fewer than half that many hearts were available.

He spent three nail-biting weeks waiting. Once again, he received a life-changing call in the middle of the night. Only this time, it was finally good news. It was his heart surgeon, who told him a hepatitis C–infected heart had become available. Without hesitation, he said, "Let's do it." Within eight hours of that call, he had his heart transplant. He did contract the dreaded and dangerous hepatitis C infection as expected, but his own experimental protocol thankfully worked and he was cured of the infection two months later. As a result of his sacrifice and willingness to test his protocol on himself, hepatitis C–infected organs are now routinely offered to everyone, and other transplant centers have followed his lead.

Today, Dr. Montgomery is almost six years post-transplant. I admire him greatly for what he has gone through and what he has contributed to science, medicine, and society. Many people will come to owe their lives to him. Most of all, I have deep admiration for the ethical and moral challenges he has faced and overcome, especially his willingness, first, not to prioritize himself on the transplant list when he probably could have, and, second, to experiment on himself with a hepatitis C–infected organ knowing that he might potentially contract a deadly infection. This decision highlights the crux of the message from the recalled experience of death: *It is possible to be fully active in life, yet at the same time live every moment with a deeper sense of purpose and meaning based on universal moral and ethical principles. It is also possible to be more selfless and act toward others as one would want to be treated, no matter how challenging the situation.*

Today, Dr. Montgomery is now back to work, operating, traveling, enjoying the outdoors, and giving lectures. He writes: "My heart disease is gone; gone when my heart was removed. Gone also is my ICD. I am no longer at risk for a sudden cardiac death. I traded all that for the life of a transplant patient, which has its own set of challenges, but I clearly traded up. It is cliché, I know, but I enjoy every moment of every day. It is a miracle I am still here."

When discussing his seven encounters with death, Bob did not describe his consciousness as diminishing or fading away in some way; in fact, he had experienced the complete opposite. He said, "The experience I have had in these sudden deaths *has always been exactly the same.*" He continued to explain that his consciousness had suddenly felt immensely vaster than anything he either had experienced or could have imagined before. This is how disinhibition in relation to death seems to open access to new dimensions of reality for people. It was as if during ordinary life his consciousness had been held back and heavily constrained by some sort of metaphorical braking system, yet after biologically crossing into death, those brakes had suddenly been released, giving him access to what had been experienced as *his own underlying immense* and *ever-present* consciousness. In short, it had felt like his own ordinary consciousness had expanded by multiple orders of magnitude into what could well be called a *state of vast hyperconsciousness.*

The first key feature we discovered through our research, which hasn't been widely discussed, is that people report that *with death, the ordinary state of consciousness becomes enormously expansive, and with that comes a much higher sense of reality*. He explained that "we are not [ordinarily] aware of that because it's just too vast to imagine, and that's how it felt." Furthermore, "it felt so permanent" and "it was always the same every time [during each of the seven times that his heart had stopped]." He added, "It didn't matter how long I was out, or what had happened in between . . . and sometimes a lot had happened. There was no deviation from that story each time."

Dr. Montgomery, like many other people whose experiences we have come to study in the past twenty-five years, was in the process of death each time, and from the perspective of people observing him, he was in a coma: *unconscious, lifeless, limp, and unresponsive*, as they attempted to revive him. Yet his experience was completely different. He had experienced that his consciousness—*his selfhood*—was of a "transcendent" nature and had "always been there and was so permanent, like my essence." Because "it's not earthly. It has nothing to do with the here and now. It was there before, and it was there as I had my last flickers of life. It's just not something we're in touch with, I think because it's too much to imagine. *It's only when everything else is taken away* that I was able to become aware of that. It's profound."

Why is this recognition of a vastness of consciousness only revealed to some people with death and not under ordinary, day-to-day circumstances? Again, this goes back to the process of disinhibition. It is not necessary to have access to the totality of our consciousness when we are getting on with our ordinary lives, but it becomes important and critical as we traverse into death.

Dr. Montgomery explained that his seemingly vast and real consciousness, which he had experienced with death, would become "drowned out" and collapse into a greatly shrunken and limited state again every time he found himself back in his body after being successfully resuscitated back to life. This is why what he and others describe feels almost like a valve had opened and closed again. He said it felt "as if my whole earthly experience comes rushing back into this body vessel and refills it, and that transcendent consciousness starts to get drowned out and disappears into the background."

When comparing the "vastness" of his consciousness during his recalled experience of death to the limited sense of consciousness that he and the rest of us experience during our daily lives, he explained: "I would say who I am on this earth, Robert Montgomery, is in a way like the [size of the] earth, and this other [real and transcendent] consciousness [by contrast] is like the [size of the] cosmos."

This was a very revealing comparison. The size of the Earth relative to the rest of the cosmos—with hundreds of billions of galaxies, each with hundreds of billions of stars and planets—is like comparing the size of a speck of dust to our planet. This gives some perspective regarding the enormity and vastness of the sense of "hyperconsciousness" that people report experiencing at death compared with their ordinary, day-to-day consciousness. It explains why survivors often say that after returning to life, they feel their ordinary consciousness is limited and constrained compared with what they had experienced. They do not experience their consciousness as diminishing or fading away with death; instead, they experience it—*including their selfhood*—as continuing in a form that is *much more immense, vast, lucid, real, and clear.*

This state of lucid hyperconsciousness is an inner experience that occurs when people are seemingly unconscious and unresponsive, while undergoing a life-threatening illness or injury, from the perspective of others who may be observing and even medically trying to save them.

Now, of course, this is Dr. Bob Montgomery, a world-respected doctor and scientist. He does not agree with how Dr. Borjigin and some other doctors and scientists have historically tried to dismiss such experiences by characterizing them as something "unreal" and "imaginary" that arises as a trick of a dying brain. He certainly does not shy away or dismiss his experience as an artifact of an electrical current or seizure.

He is one of the estimated hundreds of millions of people currently living with these profound experiences. Multiple surveys suggest that, globally, approximately 10 percent of adults are living with these recalled experiences surrounding death. By contrast, according to a recent World Health Organization factsheet, around 5 percent of the adult population live with major depressive disorder. Assuming a population of 8 billion people globally today, and after removing the population of children, there may be around 500 million people living with recalled experiences surrounding death. While anecdotes can perhaps be ignored by science, the voices of hundreds of millions of people cannot. Thus, Dr. Montgomery's experience is certainly not an oddity or an anecdote.

Survivors with recalled experiences surrounding death, such as Dr. Montgomery, have consistently and unanimously continued to assert the reality of their experience while in turn dismissing those who try to characterize their experiences as imaginary. Dr. Montgomery further explained, "I would say I only had a sense of awareness of this state of consciousness that is not connected with who I am right now. But it wasn't like there was nothing. There was something profound, it just wasn't anything I could understand."

He had no sense of what he called "earthly emotions, fear, dread, nothing like that." Rather, he said, "it just feels like permanence and something really big." And after he recovered and returned to being conscious, he not only remembered what he experienced, but was changed by it. Each time, he had a similar experience. He says, "For the first few days after a sudden death, I have this incredible elation . . . I just felt so positive, and confident. Not at all what you would expect to follow an experience where you just died and got revived."

It is truly remarkable that Dr. Montgomery is "still here." He is an extraordinary example of the newly discovered concept of what it means to enter into the early phase of death and come back again with memories of a vast consciousness in death—feeling utterly and totally connected to everything, everywhere, all at once. Likewise, others who have entered the grey zone describe similar experiences and report their consciousnesses separating from their bodies.

Throughout this book we have—and will—discuss the effect of a recalled experience of death on patients and people who return from the grey zone of death. However, there is another group of people deeply and intimately affected by these experiences: the health care providers who bring these people back from that grey zone, tend to their bodies, restore their heartbeats and brain waves, and then must attempt to analyze and understand what has happened to their patients physically but also mentally and emotionally. Many doctors and nurses are hesitant to share the details of what they've experienced, especially when they are new to medicine and on their first rotations in a hospital. There is good reason for this: even now they are unlikely to be believed. Their experience may be discounted as inexperience, a poor grasp of medical practices, or an overactive imagination. Likewise, survivors with a recalled experience of death may be hesitant to share in the first place, for those same reasons. This is slowly starting to change as more and more doctors and nurses speak up about what they have seen.

Chapter 8

AWESTRUCK: THE DOCTORS' PERSPECTIVE

Bob Montgomery is not the only physician who has talked about a recalled experience of death. Go to any hospital after hours, when there is a lull, and if enough people are standing around long enough, you will hear stories of the inexplicable and unexpected. I've heard many, often from very well-regarded doctors and nurses. Dr. Tom Aufderheide is a world-renowned professor and internationally recognized leader in the field of emergency cardiac care and past chair of the prestigious annual American Heart Association Resuscitation Science Symposium.

He and I met for the first time in 2012 after I had been invited to share my work at the Emergency Cardiovascular Care Update (ECCU) conference. The audience comprised world leaders in resuscitation science and representatives from device-manufacturing company conglomerates such as Phillips, Zoll, and Medtronic, as well as hundreds of other doctors, nurses, scientists, emergency providers, community advocates, and, importantly, survivors of cardiac arrest.

At the end of my presentation, Dr. Aufderheide stood up and shared a compelling story with the audience. He recounted, "On my first day as a physician, I was on call, and I was justifiably terrified, because the day before I had been a lay person and now I was suddenly a physician."

Doctors all find their first day at work to be nerve-racking. On that first day of his first year as a resident (or trainee doctor), Tom was assigned to work alongside a slightly more experienced, second-year resident. This resident had promised Tom he would cover his back and "be present throughout the entire upcoming 36-hour grueling on-call shift." This made Tom feel more secure, as he might have to deal with a sudden life-and-death medical emergency at any time.

"We had been up for well over twenty-four hours," Tom recalled. "It was about four o'clock in the morning, and he said that he was tired and was just going to go to bed to get some rest. He told me there was a new patient up in the coronary care unit and that I should just go see that patient on my own. He would join me later, at eight o'clock in the morning."

Tom found the new patient and introduced himself. He said, "Then suddenly, the patient's eyes rolled up in the back of his head and he fell back onto the bed lifeless." He continued to explain, "Well, here I was, frankly with my worst nightmare. This was my first day on the job as a doctor and absolutely the worst thing that could possibly have happened had just happened. To top it all off my second-year resident, who had promised he would be there to help, was absent."

A fleeting thought rushed through Tom's mind: *How could you possibly have done this to me? You have left me alone with my worst nightmare.* He immediately pulled himself out of that state of self-pity, as his instincts and medical training "just kicked in," as he put it. He realized he was not only the one doctor in the room, but also the only person who could save this man's life.

"I had to do something," he explained. "So I dove on his chest. I started CPR, and then I yelled for the nurse to help. She brought in [an emergency medical] cart, and I defibrillated him [gave him an electrical shock treatment] and initiated resuscitation practice [this means giving round after round of rapid chest compressions and electrical shock treatment, together with epinephrine, a powerful blood pressure–enhancing drug, and breaths with oxygen]."

This continued for "quite a while." The patient "would respond to an electrical shock and his heart would restart again. Then it would stop again, and he would go back into cardiac arrest." The man was suffering with something resembling the dreaded ventricular tachycardic storm, the same condition that Dr. Robert Montgomery had barely survived in Italy.

Tom recounted, "I kept shocking this person over and over, and he kept getting a pulse back and then re-arresting." By now, the hospital cardiac arrest (code) team, including his supervising second-year resident, had also showed up. He said, "This continued for an hour, and finally they said, 'We have other things to do [patients to see]. We will leave you to shock the patient when his heart stops, and if his heart doesn't restart, let us know and we'll return.'"

Exhausted, Tom sat next to the patient, and within "five to ten minutes" the man had rearrested. He said, "I would shock. He would get a pulse back before losing it again within minutes, and this cycle continued throughout the whole morning." The patient remained unconscious and in a coma throughout and was medically unstable while hovering between life and death, on the coronary care unit, attached to life-support systems.

"Eventually his wife showed up and the nurse came in and said, 'Will you please talk with his wife? I'll shock the patient if he rearrests.' So, I went out and talked to her and I was not very optimistic at all about his chances of survival. Then, at about 11:30 a.m., his lunch showed up and since he was unconscious and I hadn't had breakfast or lunch, I asked the nurse, 'Is it okay if I have his lunch?'"

"She said, 'Well, of course, go ahead and eat.' So [in the few minutes while I waited for his next cardiac arrest] I went ahead and ate his lunch." Meanwhile, the "on and off" cycle of cardiac arrest and shocks "kept going on over and over and over again, until finally about 1:00 or 1:30 in the afternoon, the patient stabilized, had a return of pulse, and that pulse was stable." Although he had stopped going in and out of cardiac arrest, he remained critically ill while unconscious on life support for days. Tom remembered, "The patient had a very rocky course and was in the hospital for about a month."

On the patient's last day in the hospital, Tom walked into his room. The patient asked Tom to sit down and shut the door because he had a story he wanted to share. He said, "I had an experience during my cardiac arrest that has been bothering me for a month and I need your help with it." He told Tom that he had experienced leaving his body and then proceeded to explain everything that Tom had done, including how Tom was performing CPR, and the shocks. Tom added, "He said he had been able to follow me down the hallway to my discussion with his wife." Then he told Tom, "You know, I thought you were pretty pessimistic about my options, and you should have been more optimistic, Dr. Aufderheide. *And if that wasn't bad enough, you also ate my lunch.*"

Tom said, "All of that was an insight for me. But then he said something that really got my attention." The patient said, "I thought it was unusual that here I was dying in front of you, and you were feeling sorry for yourself because your second-year resident had left you alone and wasn't supervising you." Tom was unequivocal. "Now I had never verbalized that embarrassing thought to anyone, yet somehow he had managed to tell me what I had thought and that really got my attention. To this day, I have no scientific way to understand or explain how that could have occurred."

Intrigued, Tom became more curious. The man asked him to "help him process this event." But since Tom had never heard of such a thing, he suggested, "Why don't you talk with your priest or minister?"

The man replied, "Well, I did, but halfway through my story he [the priest] got up and said that's not in the Bible and walked out and slammed the door on me." Perplexed, Tom reassured him that he would try to help. "I don't know much about this, but I will try and find out for you." He called the hospital psychiatric crisis line and talked with the psychiatrist. However, "halfway through the story, the psychiatrist said, 'This is a bunch of garbage,' and just hung up."

Tom could not find anyone to help. This was in part because what his patient had described would not fit into traditional views held by either the priest, representing religion, or the psychiatrist, representing science. *In fact, recalled experiences surrounding death do not fit with any of the traditional views of what happens at death.* This is an important point, which makes these experiences stand out.

This man's predicament also highlighted how others, including doctors and scientists, but also people's family, friends, and (at least in this man's case) even the clergy, typically reject their testimonies outright, instead of maybe adjusting their own preconceived worldview to try to learn more about what might happen at death.

What this man had described to Tom also illustrates how lonely survivors of life-threatening situations with recalled experiences surrounding death can feel, and how hard it is for them to find other people who might be willing to listen, let alone help them.

Tom said his patient did not tell him just about what Tom did at the scene. He also described going down a tunnel to a light and experiencing a beautiful scene in which he was in a meadow. He communicated with his relatives who had passed away, as well as a higher being. Eventually, he was told that he needed to return.

Although this was a new experience for Tom, he said he now recognizes that this "is a well-described phenomenon of consciousness." He explained, "I didn't know what this event was at that time, and I could not help my patient. Unfortunately, he passed away before I ever learned about what he had experienced."

Throughout the rest of his career, Tom asked other survivors of cardiac arrest about their experiences, and many of them described a similar experience. He told me, "Many patients stated that while I was caring for them in a state that met the clinical definition of death—no heartbeat, no pulse, no breathing, fixed dilated pupils—yet they responded afterward, by telling me about how they were watching me and could report everything that occurred during that period of time." Ultimately, for Dr. Aufderheide, this "remarkable phenomenon seems to indicate consciousness continues irrespective of crossing into death," as he put it.

The story that Dr. Tom Aufderheide recounted is certainly astonishing and hard to explain using our current scientific models. However, this is by no means the only case of a frontline, senior doctor or other health care professional recalling how they had been left baffled by their patients' testimonies of maintaining full lucid awareness when encountering death.

Very early on in my career, when I was starting my research, I wrote a letter to Dr. Douglas Chamberlain, a famous and eminent professor of cardiology, who had for many years played a leading role in research and had helped establish worldwide guidelines on cardiac arrest resuscitation management for doctors. Although I hadn't expected a reply, to my surprise, a week later I received a letter back. In this letter, Dr. Chamberlain encouraged me to follow my plans to study this subject and then proceeded to write about a man he had resuscitated many years earlier while working as a British army doctor in 1960. This was around the time when modern resuscitation techniques were being developed.

He described that one day, after prescribing a drip to one of his patients on the ward, he had gone to the outpatient clinic, but had then been called back in a rush by the head nurse, as his patient's heart had stopped. He explained, "On my arrival the patient did not have a pulse, so I started resuscitation. Techniques were more primitive then, but I persevered and then continued until eventually the patient's heart restarted. The man eventually made a full recovery and then started to tell me that during the resuscitation process, he had been at the ceiling watching me working on him. He also went on to tell me that without wanting to sound ungrateful, he really didn't want to come back!"

Soon after this episode, around 1998, I was working the on-call shift with Dr. Richard Mansfield, an experienced and meticulous cardiologist at Southampton General Hospital in England. While walking down the corridor and talking about routine things, he suddenly pulled me aside and said, "There is something I want to tell you." He paused before continuing. "I have never told anyone about this, and I don't ever want to talk about it again. This is something that really freaked me out and I still cannot explain how it happened, but it did. In fact, if you ever bring this up again, I will probably deny it, as I really do not want to talk about it. I have tried to block it out of my mind."

He shared how one night when he was working on call at Leeds University Hospital, in the north of England, there was a cardiac arrest. His pager had gone off, and, together with the rest of the cardiac arrest team, he had run to find a thirty-two-year-old man without a pulse and who wasn't breathing. He was in asystole, meaning his heart had stopped and flatlined. Richard, as the most senior doctor, was the cardiac arrest team leader. He and the team administered CPR. He explained, "We had intubated [placed a breathing tube in] him and he was receiving oxygen, as well as regular cycles of chest compressions and adrenaline [epinephrine]. We also gave him atropine." Despite all efforts, the man had no heartbeat, and his heart remained in asystole (flatlined). Richard continued to check for a pulse regularly—in fact, every two to three minutes—but throughout this whole period, the man never regained a pulse and his heart never started again.

"We carried on for over half an hour and had lost all hope of being able to save him, as he was pulseless and in asystole throughout this period," Richard explained. Nonetheless, because he was young, the team had decided to continue even though the chance of bringing him to life was almost nonexistent. He continued, "After close to an hour, it was obvious we were not going to succeed. So, as the cardiac arrest team leader, I discussed with everyone else, and as a team, we decided to stop CPR and declare him dead." But before stopping, out of an abundance of caution, Richard made a point to again check the monitor and leads to reconfirm that the flatline state was real (even though he had already checked this many times). "He still had no pulse. So we stopped and declared him dead . . . We were all very sad about this, as he was so young."

The nurses called the patient's family to tell them their son had died, and they started to clean and prepare the body, so he would look more presentable for his family's arrival. Richard went outside the room and started writing about the event in the medical records. Then Richard realized that he couldn't remember exactly how many vials of adrenaline (epinephrine) he had used in the cardiac arrest. So about fifteen minutes after they had declared him dead, he went back in the room to check how much adrenaline they had given. "While I was standing at the head of the bed and talking to the Sister [in the United Kingdom, head nurses are often called Sister]," Richard recalled, "I looked across at the man and noticed that he wasn't quite as blue as when I had left him. In fact, he definitely looked pinker, which was very strange." Richard looked at him again and again. "He definitely looked pinker than when we had ended the cardiac arrest," he said. The head nurse agreed. "I think you are right; he doesn't look as blue as before." Rather hesitantly, Richard leaned over and put his hand in the patient's groin to check for a pulse. He said, "Unbelievably and to my complete shock, the man now had a pulse. I couldn't understand how, but he definitely had a pulse." Baffled, he and the head nurse looked at each other and immediately resumed CPR and called the cardiac arrest team. They eventually stabilized the man and transferred him to the intensive care unit on full life support. Just as had happened with Tom's patient, this man had also been in a coma and completely nonresponsive throughout.

Richard recounted that about a week or so later, while rounding on the hospital ward, he unexpectedly came across that same young man. He said, "To my amazement, not only had he recovered fully but he hadn't suffered any brain damage either. Everyone on the cardiac arrest team had been sure he would have suffered severe brain damage. He had been through a prolonged resuscitation [this is because patients are oxygen deprived, as CPR cannot deliver sufficient oxygen to the brain, even when performed really well], but had also been left for around fifteen minutes afterwards without any kind of treatment or oxygen whatsoever."

Richard continued, "When I saw him later, he told me that he had watched everything we had done from above and described it all in detail. This included everything I had said and done, including how I had been checking his groin for a pulse, the details of our conversations, as well as how I had decided to stop resuscitation and had gone out of the room, before returning again. He also described how we had restarted the resuscitation efforts and how the cardiac arrest team had returned. He got all the details exactly right, which was impossible because he had been unconscious and in a coma throughout this whole period while in asystole [flatlined] without a heartbeat and dead. Then for the last fifteen minutes or so, we had left him as a dead body without any CPR, while we were cleaning him and waiting for his family to come in."

Some people might wonder whether those individuals had watched scenes of cardiac arrest resuscitation on television and later absorbed those memories and interpreted them as their own experiences. However, people often describe very specific details that relate to their own specific case, which sometimes leaves their doctors totally baffled. A dream or imagined experience would certainly not be like this. Even the man in Dr. van Lommel's study knew exactly where his dentures had been placed, and the man in our AWARE-I study recalled being shocked exactly twice, which his medical records verified. He could have guessed any number of shocks.

Richard continued, "This is why to this day I didn't tell anyone else . . . I also know that I checked the monitor, the leads, the gain [this is a technical means of checking the monitoring apparatus], and the connections, as well as the pulse, before stopping. I just can't explain it and that is why I tried not to think about it anymore and that is why I really freaked out."

I knew and trusted Richard Mansfield; not only was he a competent cardiologist, but also an excellent and methodical doctor. Furthermore, he had no interest in this subject whatsoever, yet here he was, like Douglas Chamberlain, having faced something extraordinary and inexplicable that he did not believe could have ever happened.

Many other doctors and nurses who work with critically ill patients also recount similar testimonies. One came from an American doctor working in the intensive care unit at the world-famous Great Ormond Street Children's Hospital in London. She had trained in pediatrics at Harvard Medical School before moving to England to gain further experience in intensive care medicine. After finding out about my interest, she started to tell me about a critically ill child she had cared for a few years earlier. She said, "At that time, I was part of a team, and we would go to smaller hospitals [with fewer resources] and collect their sick children who needed to be transferred by ambulance to Great Ormond Street Hospital for specialist treatment. One evening I had gone with the rest of the team to a hospital in Kent, about twenty miles away from London, to collect a nine-year-old child with severe kidney failure. She was very ill and needed urgent transfer to our pediatric intensive care unit. During the ambulance journey, we got stuck in the rush-hour traffic, and even though we were going as fast as we could with the emergency lights and sirens blazing, we could not travel fast enough. The girl's condition deteriorated and suddenly her heart stopped beating and she was in cardiac arrest. We started resuscitating her straight away in the ambulance. We tried over and over again, but just couldn't start her heart again.

"Eventually one of the nurses said, 'Look, she is dead, why don't we drive off the main road and go to a local hospital and have her pronounced dead?' [By law, ambulance staff could not declare someone dead. It had to be in a hospital.] Something in me said we should carry on, even though it looked as though we really had lost her. I said, no, we will carry on with the resuscitation. If she is going to be pronounced dead, it will be at Great Ormond Street and nowhere else. So, we carried on with the resuscitation. I didn't have much hope, but something told me to start talking to the unconscious child during the resuscitation. I don't know why, but I did it, even though it made no sense to me really. Nevertheless, I kept comforting her and telling her not to worry and that she was going to be okay.

"Amazingly, we got her heart restarted [after about 45 minutes] almost as we arrived at Great Ormond Street. Although she was still in a very critical and unstable condition, at least we had got her to Great Ormond Street." She explained, "I never looked after her [the child] again, but I heard that she had gradually improved and eventually was discharged home.

"Many months later, she [the child] came back to the hospital with her parents to see everybody who had cared for her. During her visit she asked one of the nurses, 'Where is the American doctor who looked after me in the ambulance and who was talking to me during the trip?' She then described how she had watched everything we had done from above and had recalled all the details. I was amazed when I heard this, as she had never even seen me throughout the trip. She had been too ill and unconscious throughout, while on life support . . . "

All of these patients, from Dr. Bob Montgomery to the little girl in the ambulance, had all traversed into the grey zone of death. Dr. Montgomery was in cardiac arrest, not just once but seven times. Dr. Tom Aufderheide's and Dr. Richard Mansfield's patients were fully submerged, wading deep into the grey zone. The child in the ambulance also was in cardiac arrest and completely unconscious. And yet these people could describe everything that had been going on, and these events could be corroborated by others—and not just the events themselves, but even sometimes the thoughts and feelings of the doctors too. This was all happening while people had felt a sense of separation of their consciousness from their body.

The experiences of the patients are extraordinary enough. But equally noteworthy is the urge that doctors and nurses who have come across such cases feel to come forward and go on the record to share what they witnessed as their patients had drifted into the grey zone of death. Like Dr. Aufderheide found, most patients still find it hard to get the assistance they need. His case raised difficult, complicated questions that no one at that time had the insights or evidence to answer. It was bad enough that the patients didn't get the help and support they needed, like Aufderheide's patient, who needed someone—anyone—to take his story seriously and help him understand what had happened. This is slowly changing. I think that if that unfortunate man had asked for help now, he might have at least had a friendly, if disbelieving, ear. We hope, as more and more individuals come forward to share their experiences, that the stigma of a recalled experience of death will fade further, for both medical professional and patient alike.

Chapter 9

LIBERATED AND SEPARATED

ONE OF THE MOST INTRIGUING OBSERVATIONS ABOUT THE recalled experience of death is how different the experience is for the person dying and the people desperately trying to save him or her from death. People undergoing a life-threatening illness or accident may naturally experience a tremendous sense of panic, fear, and distress as their condition deteriorates and they fight for their life. Then suddenly everything goes blank, as they lose consciousness from the outsiders' perspective. This person who had been teetering around death appears unconscious and seemingly dead, yet *from the perspective of the person who is going through death, there is an inexplicable sense of continuation of consciousness, neither interrupted nor halted.* Their consciousness now feels more lucid, with visual and auditory awareness regarding events. At the same time, the people on the "outside," the doctors and nurses, are fighting to revive the individual. There is blood, shouting, calls for more equipment. Perhaps weeping loved ones in a corner.

This inner perception is experienced as a separation of consciousness and selfhood combined with a vast expansion of consciousness, which is very much felt as a release and liberation from the body. This is all experienced without pain or distress, even if the person is undergoing seemingly painful lifesaving medical procedures. This individual—wading deeper and deeper in the grey zone—may look at this frenetic activity with equanimity. What happens to this body on the gurney seems unimportant.

Based on our research, this phase of the recalled experience of death, which others often refer to as "out of body," can be divided into at least fifteen distinct components, or subthemes:

1. *Vast expansion of consciousness.*
2. *Separation from the body with external visual awareness.*
3. *Liberation and weightlessness.*
4. *Hovering or floating in the space around the body.*
5. *Observing the body or events from above.*
6. *Becoming detached from events and the body.*
7. *Realization of having died.*
8. *Lucidity with reasoning and clarity of thought.*
9. *Initial confusion. This is often transient and often goes away after the person becomes accustomed to their new situation.*
10. *Taking on the appearance of a transparent humanlike form or a light.*
11. *Bird's-eye view—in all directions—as if through 360 degrees.*
12. *Consciousness pervades everything.*
13. *Shedding the body.*
14. *Connection by a metaphorical cord.*
15. *Selfhood and consciousness: not the same as the body.*

Some of these subthemes have been described by other researchers, too. However, some components—most notably (a) the perception of remaining connected to the body by a cord, (b) selfhood and consciousness pervading through everything, (c) taking on the appearance of a transparent human form or a personified light, and (d) the realization of the enormous magnitude of one's own consciousness (analogous to the cosmos compared with the Earth, as Dr. Montgomery had described)—are probably much less familiar to people.

To paint as complete a picture as possible of this stage of the experience, these themes are presented together. I have tried my best to divide the subthemes—to make it easier for readers to differentiate them—but since people describe them together, there will be some inevitable overlap in the following exemplary testimonies.

1. VAST EXPANSION OF CONSCIOUSNESS

Dr. Montgomery's experience highlighted the expansive consciousness experienced in death. Many others have described something similar. Bolette, a fifty-two-year-old Danish woman who describes herself as having been a politically active nonreligious woman when she suffered cardiac arrest due to a complication of childbirth at the age of twenty-six, recalled her sense of heightened lucidity as follows: "I felt myself to be very awake and aware the whole time. I was immensely curious and observing, and *my awareness was . . . much larger than when I am here in life.*"

This sense of an expansion in consciousness is a feature throughout the whole of the recalled experience of death. This includes during the sense of separation from the body. It may be described in different ways by different people, including as a sense of having sudden access to tremendous levels of knowledge. Joseph, a man in his seventies, who had experienced a climbing accident as a teenager after a large rock had broken loose and fallen on him in 1966, explained, "The next thing I knew I was floating there, hovering in the air . . . *Instantly my head was flooded with knowledge . . . I couldn't believe how clear my thoughts were . . .* "

Renee, an American motorcyclist, had a severe accident in 2003. Her heart stopped, but she was revived back to life and was placed on life support. She explained her experience as follows: "I could see myself outside of my body. My heart was lying outside of my chest, and it was not beating. Strangely, just like watching a big-screen movie, I could hear everything going on around me. But I was dead. How could this be? I could hear the police and the paramedics talking about needing to resuscitate me, and how much blood I had lost."

She then explained, "My emotions are heightened [and I] accelerated into pure light energy where there is no space or time. I felt more alive than I have ever felt in my life. There are no words to describe this ascended transformation. It is not humanly conceivable to understand. If you can, imagine all of a sudden . . . I became this all-knowing, all-powerful, amazing, incredible light of energy."

Michael, an American man who had suffered a cardiac arrest during heart surgery in the early 1970s, likewise explained, "I felt like I 'knew' everything for just an instant." Another American, Karen, explained, "Before I died, I questioned everything. Here [in this new state during her experience], I knew everything and there were no more questions."

In 2014, Jeremiah suffered a cardiac arrest during a complication with surgery. He explained, "My speed of thought and cognitive abilities became extreme . . . Maybe it was tenfold in capabilities. I immediately 'knew' a whole lot of information. For every thought, the answer was immediate." During her recalled experience of death, Krista, an American physician's assistant who suffered a cardiac arrest that lasted around eight minutes during childbirth, explained that "I was simultaneously given the answers to all the questions I'd ever had in my life."

Sharon shared, "I began feeling as if I was attached to a giant IV bottle of knowledge. I was being fed all this knowledge . . . I felt such joy and elation; it was one 'A-ha' moment after another. And it all seemed so simple and so logical."

Katherine recalled, "The way I experienced knowledge and information wasn't through human means. I experienced a sort of fused knowledge, where I had access to different aspects of knowledge if I focused on whatever it was I wanted to know. In this way, I'd immediately know the answer to what concerned me." Telesa explained, "I understood things that I couldn't possibly have known." Patsy, another American woman, explained, "Suddenly, I had all knowledge." And Rachel, a British mother of four, whose testimony we will read in far more detail in chapter 10, had simply said, "I understood everything."

2. SEPARATION FROM THE BODY WITH EXTERNAL VISUAL AWARENESS

Ingrid, a fifty-year-old woman from Colombia with a PhD, described how, as a three-year-old child, she had fallen into a reservoir of water and was drowning. She had been in great distress, before feeling a sensation of being released. She said, "Then suddenly came a feeling of total relief. I didn't know why then, but I was at peace and there was no anguish. My heart stopped pounding; everything turned absolutely still; I was no longer cold and didn't feel like fighting anymore. *My body simply became insubstantial* . . . I noticed a body suspended in the water . . . This was my own lifeless body, but I was neither surprised nor frightened to recognize it. Instead, I felt an immense joy and freedom . . . So, I simply kept going away without ever looking back at that body."

People routinely describe their consciousness and selfhood as ejecting from the body, which leaves them with a tremendous sense of freedom and happiness. Another person explained: "Suddenly, I was above my body, which lay on a stretcher, wearing a white hospital gown. *I looked at my body and knew it wasn't the real me, it was the thing I had been caught inside, and now I was free!* Oh and how I felt such happiness!" People recognize, as Dr. Montgomery had explained in his testimony, that their selfhood transcends their body. It is as if they had been trapped in their body throughout ordinary life, which snuffles the enormity of their consciousness into a small sliver. But in this liberated state, people now perceive the vastness of their own consciousness.

3. LIBERATION AND WEIGHTLESSNESS

Ana Cecilia, a young mother with a life-threatening infection, recounted being on a ventilator in intensive care when the breathing tube became blocked. The doctors tried and failed to get her air. She says, "The dizziness started, and I began to lose consciousness. I stopped breathing and almost immediately my heart stopped. Then, I left. Suddenly . . . *I felt totally liberated,* I saw myself in a hospital gown, with the doctors around my body trying to resuscitate me. I saw how they were busily moving from one place to another, each time smaller and more distant . . . *I floated alone. The relief was enormous.* I couldn't control that which had happened, I let go. Then, there began the most wonderful journey."

4. HOVERING OR FLOATING IN THE SPACE AROUND THEIR BODY

People describe hovering or floating and observing with curiosity whatever the doctors were doing around their bodies. One person explained: "Suddenly it all happened so fast. I could see myself outside of my body. *[I] was above my body floating around.*" Stefania, a sixty-year-old Italian woman who had undergone heart surgery at the age of sixteen for congenital heart disease, unexpectedly suffered a complication in the postoperative period. While unconscious, she says, "I was no longer in my body. *I floated without weight or physicality.* I was above my body and directly below the ceiling of the intensive therapy room." Someone else explained, "The first thing I experienced was that I could suddenly see my body from above. I saw that the doctors were incredibly busy treating me . . . I felt wonderful and light where I was. I had no pain and no problems . . . *I clearly remember that I hovered over the doctor who conducted the treatment.*"

5. OBSERVING THE BODY OR EVENTS FROM ABOVE

People do not feel frightened or alarmed by seeing their own body in this state. Often, they find it intriguing to observe medical staff trying to save them. Stefania went on to explain, "I felt very good where I was now and was very surprised at the spectacle that was happening around the body which had belonged to me. I looked on with my whole being. *I was having the experience from a state of consciousness which completely pervaded my being* and I perceived everything from that vantage point. I 'knew' that the doctors were making a big deal to get me back from where I was. I 'knew' that other surgeons had arrived from other wards to help out, *but I did not want to return from where I was.*"

Another woman, Deborah, had been rendered unconscious after suffering complications, including severe blood loss, after a ruptured blood vessel in her abdomen. She explained, "All I know is that I was floating, upward for a while, then just forward and there was so much to look at . . . [I could] clearly see myself lying on a table surrounded by seven doctors. I was being operated on. The only part of me that could be seen was my closed eyes, some bangs and my heart feverishly being repaired."

Susan, an American woman who had suffered a cardiac arrest in 2013 after a severe allergic reaction, explained: "I saw the [medical] team enter and surround a person lying next to me; at least that was what I thought at the time. It never dawned on me that the other person *was me*. I was very, very calm. But I felt bad for the lady (me) as the code blue team [cardiac arrest emergency team] struggled to get her heart to start again. I saw 'me' being intubated [having a breathing tube being inserted] with chest compressions and the rotation of the team to revive me."

6. BECOMING DETACHED FROM EVENTS AND THE BODY

The pull of this new and powerful experience is such that people become mentally detached from their day-to-day worries and lives, even their families and their body. Stefania recalled, "I, who no longer was the body that had belonged to me just a moment prior, found myself in a position which was . . . more elevated." She continued, "As I was floating free without physical limitations, I experienced a feeling of infinite, supreme happiness. At the same time, I experienced an extremely alive and vigilant state of mind. I understood at a very deep level what was happening [to] the body that had belonged to me and was now intubated [had a breathing tube and was on a ventilator] on the hospital bed in intensive care. *The thought was very lucid and came through immediate knowing.* It was a different way of thinking . . . *I recognized the body as mine but I was no longer interested: I was not that body.* From this dimension, I observed everything that was happening and all the succeeding events without being involved."

Krista further explained: "[I] was now floating above my body. The distance between my conscious self and my body below seemed to be stretched out . . . There was no sound, no pain, and no fear . . . I could see someone lying on a bed . . . As I lingered above, I didn't identify, in any way, with the body or the people in the room. I was instead a detached observer, although still 'Krista' within the fine static of my consciousness. *I . . . felt more alive than ever.*"

Karen recalled how she had suffered a ruptured appendix with gangrene and was bleeding so heavily during surgery that she suffered a cardiac arrest. She explained, "I looked down and saw my body with many people around it. I did not feel any attachment to my body or regret upon leaving it. I felt so light and free: free of the pain of the past several weeks and free of the pain of my life up to that point."

Wendy, a Canadian woman who suffered a cardiac arrest in 1982, explained, "[I knew I] would be leaving behind a 5–6-month-old infant and my husband, but I did not care. [Because the magnetic pull was so powerfully strong that it overwhelmed even her powerful maternal instincts, and she found that she simply couldn't resist rather than that she didn't care.] I wanted to go into the light. I wanted to go home."

7. REALIZATION OF HAVING DIED

In this state, people come to realize they have died, as Moody also pointed out. Joseph, after being extracted from the climbing accident on the mountain, lost consciousness and was taken to a hospital for surgery. He recalled, "Then it hit me like a ton of bricks. *I was dead and yet here I was, still alive and fully conscious* . . . I watched as the doctor, the chief surgeon, took a saw to me to quickly open me up."

Ron, a man now in his seventies, had suffered a terrible car accident as a teenager. He said that after sinking "into unconsciousness, [then] suddenly I was *totally alert—more alert than I had been in my life—more alert than life.* I was totally free of worry and doubts and bothersome physical sensations and limitations. I was floating near the high ceiling of a room in the Breeze Community Hospital. At the time, this seemed perfectly natural and normal . . . I recognized Dr. Ketter in the room . . . *I realized then that I was dead . . . There is no experience . . . to draw a parallel [with] . . . I felt a supreme sense of peace and an absolute lack of fear.*"

In 1996, Wayne was trying to help move a broken-down car off the street when another car rammed into him and the car from behind. He said, "I was between the cars and the impact was so great, it bent the frames on both cars." He was taken to a hospital on a gurney. He felt like he was going to die, before losing consciousness. He explained, "I knew I was going to die, and welcomed the release from pain." He then said, "I came out into the light and was in an upper corner of the emergency room looking down at my body on the gurney. *I was not disturbed by being dead, or by seeing my body on the gurney.* I was in a state of euphoria and a sense of perfect peace and being. I had no pain, wants, or needs of any kind. I had a sense of being home . . . I felt myself floating in the air and looked towards the gurney where I saw my body, but I was above it . . . "

8. LUCIDITY WITH REASONING AND CLARITY OF THOUGHT

In this state, people describe the ability to maintain higher thought processes, including reasoning while hovering above their own body. Bonita, a woman with a PhD in developmental studies, went into cardiac arrest during childbirth. She said, "I suddenly realized that I was looking down on myself as I lay in bed. Being a person who tends to analyze most situations before making a conclusion about a situation, I hovered above myself . . . I realized that it is impossible for people to see themselves in three dimensions. Therefore, if I could see myself in all three dimensions at a time, I would have to be outside of my body. *I reasoned that if I were outside of my body, then I'd have to be dead.*"

9. INITIAL CONFUSION

Although there is a vast expansion of consciousness, at the same time, people may find themselves initially confused by being in a new state outside of their body. However, this is usually transient and goes away after they become accustomed to their new situation. One person explained, "I was wandering around in the dark and eventually found others who were similarly confused as to where we were and what was going on. This is hard to explain, but we didn't really have a physical presence." Joseph described the sense of confusion in the following manner: "[I was] feeling stunned, not sure where I was, who I was or even what I was."

10. TAKING ON THE APPEARANCE OF A TRANSPARENT HUMANLIKE FORM OR A LIGHT

Arshan, a man from the town of Ghonbad-e Kavus, which is situated in the northeast of Iran, said he had been hit by a car after he crossed a street intersection in 1996. His experience exemplifies many of the features that others have also described. He said, "Suddenly, I heard a loud car horn followed by a loud crashing sound. At that moment, I found myself floating in a dark space. I was outside my body, floating in the air and just looking around. I saw a body lying in the middle of the street next to a car. I was looking at it from several feet away. It took me a little while to recognize that it was my own body that I am looking at. I had no feelings for it; *I was just an indifferent observer. I thought to myself that I must have died, but I was not sad at all.* I didn't know where I was supposed to go from there. My thoughts and mind were the same as when I was in my physical life . . . After a short while, I gave up the worry of where I needed to go from here, because I was enjoying the peace and silence. I was immersed in that moment. I was just watching from several feet up in the air as people were rushing towards my body from every direction. *I couldn't hear their voices clearly, yet I was able to comprehend what they were saying. When I looked at these people, I knew their thoughts and what they were going to say.*"

Arshan was quickly loaded into an ambulance, and it sped toward the hospital. "In the ambulance, the emergency medical team injected something into my body, but it was no use. My body did not respond. *Although I was detached from my body,* I still felt like I was also somehow inside of my body too. Nevertheless, I was not feeling discomfort or pain. At the same time, I was feeling that I was going higher and higher each second."

He said, "I was floating like on a wave and felt so light. At the beginning and end of the accident, everything was moving so fast. *When I exited my body, it was in another form, that was transparent and non-physical, yet it was similar to a human form*. Although I saw the new form, I wasn't giving it any thought. I was feeling pleasantly warm, could not smell or taste and did not have any bodily physical senses. I could not feel physical things, *but my eyesight was greatly enhanced. I felt like I had turned into energy*." Arshan, like most other interview subjects, was unafraid during his experience.

During the recalled experience of death, people experience themselves and others as either a personified light or transparent form of their earthly self. One subject said, "I had left my body and began to return home. I suddenly saw myself as a light . . . I was still me, Adriana. I was not my body, but I was my essence."

It should be pointed out that in the context of a recalled experience of death, when people talk about seeing a "light," which is common, they are usually referring to experiencing luminous personified entities, of greater or lesser magnitude, which are perceived as a light. This may include themselves. Timothy, a man in his early sixties who had been involved in a roadside accident at the age of twenty-nine, explained, "The next thing I remember was standing on the hillside. Beside me was the four-wheeler turned upside down and I saw myself lying face-down underneath it. I didn't pay much attention to that scene because coming toward me . . . were two beautiful . . . lights [who] transformed into two [personified] beings."

As we shall see (in chapter 12), at other times people may also use the term *light* to describe what the place they go to feels like.

11. BIRD'S-EYE VIEW—IN ALL DIRECTIONS—AS IF THROUGH 360 DEGREES

As people find themselves hovering above or around their body, they describe a perception of being able to gather visual information from all around, as if through 360-degree vision of their entire surroundings. A young woman who had also suffered a car accident, but whose heart had stopped due to heavy internal bleeding in the operating room, said, "During surgery, I popped out of my body . . . Surgery appeared more brutal and bloody than I imagined it would look, especially from a 360-degree vantage point. I could see the doctors and the entire room all at once without blinking or relying on eyes. There, in the room with the doctors, nurses, surgical technicians, anesthesiologist, and others, I felt incredible joy and shock as I realized all does not die with the body." Someone else explained, "I was aware that I wasn't watching anymore from behind my eyes but from in the room so I could see my body on the bed, and I could see the two nurses behind me." A Canadian man with a similar experience said, "I had 360-degree vision, I could see above, below, on my right, on my left, behind, I could see everywhere at the same time!"

Mohammed, an Iranian man, explained that in 1977 he was twenty-six. He said, "I had followed a friend's suggestion and taken a job in the city of Mashad, which is in the northeast of Iran. That day I was driving home to visit my family. It was 2 a.m. and the road was dark, when I noticed a car from the other lane was in my lane; I collided head-on with that car. I was critically injured, but luckily a few minutes later, a passenger bus took me to a small hospital. I was not anesthetized and didn't go into a coma. I remember a young woman around twenty-two years old entered the room. She seemed to be inexperienced and rather new to the hospital. She seemed beautiful and I wished I could talk to her. But I was distracted by unbearable pain and all the angry thoughts that were playing in my head."

The medical staff were working on him, but his injuries and condition deteriorated, and he eventually became unconscious. He explained, "Suddenly, I felt that everything shifted. I felt a deep calm and peace engulfing me... This feeling was totally opposite to what I was feeling a few minutes ago. I was not angry anymore *and I was seeing perfection in everything in the world and around me. Now I was feeling that everything is exactly the way it should be.*

"I noticed that I am seeing like 360 degrees around [the nurse], like I have totally engulfed her... I was aware and seeing everything there as well, without any difficulty and confusion.

"I shifted my attention to that young woman again. She seemed a little different than a few minutes ago. I could see her thoughts and feelings as well. In fact, I felt that I am present in the entire hospital. I could see that she had a lot of sadness and worry about what she was seeing. She was thinking that it was so sad that this young man is dying like this. *I tried to soothe her and tell her that I am all right and that nothing is wrong with me. In fact, I've never felt so good in my life. But she kept ignoring me, like she does not see or hear me.*

"I noticed that she was staring towards a fixed point. I followed the direction she was looking and noticed that she is looking at the body of a young man who is lying on a bed. I was shocked because this man had striking similarities with me. I wondered who he was and why does he look so much like me? I even thought that maybe I have a twin brother who happens to be in the same hospital. I was so confused and bewildered. I started to think to myself, *Am I dead? Is this my body on the bed?* My God, my mom would be so devastated. She is expecting me back home tomorrow."

Mohammed's experience highlights how during the recalled experience of death, people find their consciousness pervades everything, which is how knowledge is gained, and that visual awareness does not come about through direct vision in the way that we are accustomed to in day-to-day life. Many other people's testimonies further reinforced this concept.

12. CONSCIOUSNESS PERVADES EVERYTHING

Mache, a woman in her forties, survived a life-threatening infection at the age of four. She said, "I was looking at myself. I was blue and lying on a gurney. My eyes were slits [closed]. There were people around me frantically working. Mom was crying as she looked on." In that state, *"I wasn't upset or scared. But I was a little confused.* I knew I didn't have a body. *I felt like I was part of everything and everyone. I just floated up and could see other rooms. I later detailed conversations that there is no way I could have known. I lifted out of the hospital and continued ascending. I began browsing through time. I saw events from my short life. I later detailed things that occurred before I could even talk. These were events I should not be able to remember but I do and in great detail."*

Another person recounted, "I remember 'me' being pulled away, out of the bed, but my physical body was lying limp . . . I could still hear everything but could not acknowledge or move. I can see the nurse pushing the alarm but not with my eyes. It was as if I was looking at her from somewhere else . . . I could see so many people around me frantically doing things to me... They started resuscitation. They punched my chest with an injection. I kept thinking that this should hurt but it did not... I could see everything but not from my body. I can't explain where I was, but I was not in the body."

These cases demonstrate that while people can recall being able to see and hear things happening around them, and *can even be aware of the feelings and thoughts of the people trying to save them, because their consciousness like a flux of energy is able to pervade through others*, they can no longer communicate with those people. Dr. Aufderheide's patient had described being aware of his thoughts but had also recounted being able to follow him to another room and listen to the conversation with his wife. Others, too, have described being able to freely move beyond the immediate location where their body may be and can even gain information regarding what their relatives and loved ones had been doing.

A woman in her fifties who had accidently overdosed on aspirin as a three-year-old child recalled, "As [the medical team] worked on me, I left my body and was watching from above. I could see and hear everything but could not feel what they were doing to the body below... While watching the doctors and nurses working, I also remember being able to watch what was happening at the same time in the room they had taken my mom [her mother had been taken to a different room]. I could feel my mother's hysteria and distraught, as if her emotions were my own."

Diane, a California woman who had a cardiac arrest, said, "I felt warm, love, peace and pure joy... I looked at myself for only a moment, and at my husband frantically driving to the hospital. At that same moment, I began to multi-locate and was with my mother [and the doctor]... I could tell the doctor everything he did during my arrival. I could tell him down to the minute details."

13. SHEDDING THE BODY

Aside from a relative detachment from and indifference toward their own body, people also describe a perception of shedding their body, as if taking off an old piece of clothing.

Diane described how "I simply felt like an armor had been unlocked and my real body had been released." Another person said, "I looked at my body and knew it wasn't the real me, it was the thing I had been caught inside," while another person explained, "I had shed the sense of my body very quickly."

People often analogize the perception of shedding of the body to removing an article of clothing. Lavette, a woman who had almost drowned to death as a teenager in 1977, said, "This is real [the new state] and that body there was just a coat I had been wearing." Another simply explained that "it was like taking off your coat."

14. CONNECTION BY A METAPHORICAL CORD

Some people describe a connection to their own body through a type of a "cord" or "wisp," and they recognize that if this "cord" is severed they may not be able to return into their body.

Michael said: "I suddenly 'woke up' floating on the ceiling and was looking down. It wasn't dream-like because everything was in clear detail... There were several people in green gowns, who were working around my body on the operating table. *I distinctly remember marveling at the thin, glowing, silver 'cord' leading... down to the body on the table. It was stretched [so] very thin, [that] I started to wonder what would happen if it broke.*"

Mache said, *"It felt like I was attached to a cord."* Mark, an American, described how *"I continued to merge upward with something to which I was connected... "* Anna, a Dutch woman who had suffered a life-threatening infection as a complication after her hysterectomy, described this as a *"whitish translucent cord attached at one end to the floating 'me,' and at the other end to my body on the bed. [I was instructed] to be careful not to break that cord as I was going to need it to get back into my body."* Leonard, an American man who had suffered complications of a burst appendix at the age of eight in 1963, explained, *"I discovered a long thread hanging down from my... existence. It led down to my body on the operating table."*

15. SELFHOOD AND CONSCIOUSNESS: NOT THE SAME AS THE BODY

In this state, people come to understand that their selfhood, awareness, and consciousness is separate from their body and does not become annihilated. Romy, an Australian woman who had been in a bad car accident, explained, "I was still alive, but I didn't have my body. I know for a fact that I am, that I exist. I sensed that I had left my body . . . I have died and left my body, yet I still exist." Viva, an American nurse who holds a PhD, said, "I felt [I] rose up out of my body . . . while observing medical staff suddenly going into resuscitation mode—realizing I still existed and was not afraid . . . This must be who I really am and found that 'I' was an alive entity, a being without a body (and saw that body on the table)." A man with a PhD in cognitive psychology from Hong Kong who had survived a plane crash explained, "When [I] came out of my body, I felt like a new being, leaving behind all my old senses, thoughts, and emotions."

As already explained, one of the many fascinating elements of this first phase of a recalled experience of death is that patients report the experience as painless, no matter what is happening to their bodies. One respondent shared how her violent and abusive partner assaulted her, attacking her with a pistol, which he repeatedly struck her in the head with. "I didn't feel any pain or sense of panic during my out of body experience. I remember having thoughts like 'What is he doing now?' and 'Am I going to be ok?' I felt strangely happy, like if I had been in my physical form I would have a smile on my face. I felt no fear, pain, or concern with anything else. I remember my voice saying, 'When is he going to stop?' while looking at my lifeless body. I saw him kicking me in the face, my hands were still protecting my face but my inner self had let go of the fight. My eyes were closed and I was totally unconscious."

CHILDREN'S EXPERIENCES OF SEPARATION FROM THE BODY

Although most of our interview subjects were adults, we also had reports of children who experienced at least some of these initial fifteen stages of the recalled experience of death. Two of the most intriguing cases that were sent to me in the early years of my research came from families with young children. The first was from a woman whose grandson had survived a cardiac arrest. She wrote: "[My grandson's] heart had stopped . . . There was a lot of commotion . . . They were pressing on his chest, and he was lifeless and blue . . . They put him in an ambulance and took him to hospital . . .

"One day [after he had been discharged from hospital], during the course of play, he said, 'Grandma, when I died, I saw a lady.' He was not yet three years old. I asked my daughter if anyone had mentioned anything to John about him dying and she said, 'No, absolutely not.' But over the course of the next few months, he continued to talk about his experience. It was all during the course of play and in a child's vocabulary.

"He said, 'When I was in the doctor's car, the belt came undone, and I was looking down from above.' He also said, 'When you die, it is not the end . . . a lady came to take me . . . There were also many others, who were getting new clothes, but not me, because I wasn't really dead. I was going to come back.'"

Interestingly, the boy's parents noticed that he kept on drawing the same picture over and over again. As he got older, it got more complex. When asked what the balloon was, he said, "When you die you see a bright lamp and . . . are connected by a cord."

This young child had used children's language to describe how he had experienced seeing a bright light, which he referred to as a bright lamp. He also recalled that he had felt separated from his body and watched the event from above while still somehow connected to it by what he experienced as a type of cord. These were features similar to what adults were also describing. Then a few years later, I received another captivating case. The mother of a boy had written:

"My son, then aged three and a half years old, was admitted to hospital with a heart problem . . . He had to undergo open-heart surgery . . . About two weeks after the surgery he started asking when he could go back to the beautiful sunny place with all the flowers and animals. I said, 'We'll go to the park in a few days when you are feeling better.'

"'No,' he said, 'I don't mean the park, I mean the sunny place I went to with the lady.' I asked him, 'What lady?' and he said, 'The lady that floats.' I told him I didn't know what he meant and that I must have forgotten where this sunny place was and he said, 'You didn't take me there, the lady came and got me. She held my hand and we floated up . . . You were outside when I was having my heart mended . . . It was okay, the lady looked after me, the lady loves me, it wasn't scary, it was lovely. Everything was bright and colorful [but] I wanted to come back to see you.'

"I asked him, 'When you came back, were you asleep or awake or dreaming?' and he said, 'I was awake, but I was up on the ceiling and when I looked down I was lying in a bed with my arms by my sides and doctors were doing something to my chest. Everything was really bright and I floated back down . . . '

"About a year after his operation we were watching *Children's Hospital* and a child was having heart surgery. Andrew [not her son's real name] got really excited and said, 'I had that machine' (a bypass machine). I said, 'I don't think you did.' He said, 'Yes, I did really.' 'But,' I said, 'you were asleep when you had your operation, so you wouldn't have seen any machines.' He said, 'I know I was asleep, but I could see it when I was looking down.' I said, 'If you were asleep, how could you be looking down?' He said, 'You know, I told you, when I floated up with the lady . . . '

"[One day] I showed him a photo of my mum (she had passed away) when she was my age now, and he said, 'That's her. That's the lady.'"

The content of these children's testimonies was fascinating. In both cases, the children had experienced a sensation of maintaining consciousness, while feeling as if they had separated from their bodies. They had also gone on to draw their experiences later. Importantly, children like these could not have known much about the concept of death and what happens after death, and could not have had any preconceived ideas about it. Nevertheless, they recalled similar features as adults—with some components of the same unique narrative arc, and clear, delineated stages, that are common throughout all others' recalled experiences of death.

Aside from the sense of separation that we have just reviewed, another intriguing component of the experience reflects how people relive and reappraise their life, as admiral Beaufort wrote about some two hundred years before. Through our studies, we discovered that this component, like the separation, is composed of multiple subthemes. Before providing a breakdown of these subthemes in chapter 11, I want to share the experiences of two people whose testimonies really highlight what this review feels like for a person who has headed toward and into the grey zone of death.

Chapter 10

RELIVING A RECORDING OF LIFE

IN 2013, WHILE PARTICIPATING ON A PANEL DISCUSSION AT THE NEW York Academy of Sciences, I met Dr. Mary Neal, an orthopedic surgeon and former director of spine surgery at the University of Southern California. She had undergone a remarkable recalled experience of death while drowning, which she had later described in her book, *To Heaven and Back*. She had been invited to present her experience to a mixed audience of scientists and laypeople in New York. I found her experience to be extremely profound, particularly as it encompassed so many of the features recalled by other people, too. So, a few years later, I asked her to tell me more.

She said, "My husband and I went to South America [in 1999] with friends, for a week of whitewater kayaking. This [particular] river was a little bit different than some of the rivers we paddle in the US in that it was very high flow and very high volume."

After going down a drop, or waterfall, the front end of Mary's kayak became pinned in some rocks. She said, "While I was still upright in my boat, the water completely submerged me and the boat. The weight of eight to ten feet of water and the force of the current folded my body over the front deck of the boat and absolutely held it there."

Mary was now pinned down underwater and unable to move; however, her familiarity with medical emergencies ensured that she "didn't panic." Instead, she set about trying to free herself from the boat or "freeing the boat from whatever was holding it." As a medical professional she understood the importance of "keeping a good track of time" during emergencies. But as time ticked on in what was now her own life-and-death emergency, Mary "knew the likelihood of being rescued was pretty slim."

Despite not being religious at the time, she asked "that God's will be done." She went on: "I had never thought about the death process or what it might be like, or what I might do during my own death—other than the fact that I had always been afraid of a drowning death. I thought drowning would be the most horrifying way to die. So the irony of my situation actually wasn't lost on me."

As Mary lost consciousness, she was overwhelmed with "a sensation of being held and comforted and reassured that everything was going to be fine. My husband would be fine, and my young children would be fine, regardless of whether I lived or died." She described this as a very powerful experience of being comforted, "like when you hold a young newborn and you're pouring all of your hopes and dreams and love and your very being into that baby." Now, Mary felt herself like "that baby," even though, she said, "I was still me and still had my same pragmatic, analytical sense about me. I was experiencing this incredible sense of not just calm and peace, but this incredible sense of absolute love."

As a doctor, Mary's experience of death had always been somewhat abstract—something she observed happen to other people. She had never really thought about death until it was right in front of her. To her amazement, she wasn't scared or panicking.

She said, "Then I was taken through a life review. This was like nothing I personally could have imagined." That was because "if someone had asked me about a life review [before], I probably would have come up with a Hollywood version, where you feel good about some things you do, and then some things you wish you could redo. But this review was very different."

The reason why this was different was because, during the review, Mary, like Admiral Beaufort more than two hundred years before, came to experience not just "some things," but the most minute details of her life, which she explained felt like they "sort of went across" her consciousness. This review included "the most horrible, painful events of my life." By this, she clarified, the "ones in which either I very deeply hurt someone else, or more commonly when someone else really, really deeply hurt me, or someone I loved." She felt that during this experience, her selfhood had "*reinserted within that event*," meaning she was "once again, absorbed into that event." She clarified: "Intellectually, I remembered the emotions that I'd had. I could recognize and remember my emotional responses, [meaning that] I was remembering the guilt or remorse that I felt after I'd hurt someone else. I also remembered my feelings of bitterness and anger and hurt that I had felt [in life] when someone had [either] hurt me or someone I loved." She described these as negative, destructive emotions.

During the life review, like many other people with recalled experiences of death, Mary, too, described gaining an understanding of the "cause and effect" relationships—*the chain of reasons*—behind all the events in her life, together with the longer-term downstream consequences—*the domino cascade*—of her own and other people's actions. She described how she had gained tremendous insights regarding the existence of a deeper meaning to seemingly trivial day-to-day events and also the hardships she had experienced in life. She came to recognize a greater purpose to everything in life—even though she had not previously been aware of such a purpose during her daily life.

She explained, "I simultaneously had this incredible experience of having a complete understanding and knowledge of the 'me' within [each] situation. I knew every single part of my life story . . . my backstory [meaning the reasons leading up to the events] that had brought me to that moment in time, where I hurt someone else. I understood everything about my hurts and sorrows and disappointments and everything that led up to that moment. In addition to that, I had a complete and absolute understanding and knowledge, not just about the motivations of the other people, but an absolute knowledge of their life story, their backstory, and again, everything within their life story that had brought them to that moment in time, where they either hurt me, hurt someone I loved, or I hurt them."

Mary, like many others, had experienced the *cause-and-effect relationships* behind how specific people and incidents had come into her life. *Of course, understanding the cause-and-effect relationships behind events and phenomena is the hallmark of scientific enquiry.* Yet somehow in death, the cause-and-effect relationships that underlie human interactions come to the fore and become evident and apparent to people, even though they had not understood such detailed relationships existed before, during their ordinary day-to-day lives. *This is one of the features that indicates greater insight and knowledge in people during their recalled experience of death.*

Mary recalled her negative and destructive emotions: anger, bitterness, shame, guilt, remorse. Yet as she came to understand the backstory to these emotions, their power dissipated. She explained that "what I discovered [instead] is that the only emotion that I felt was incredible, compassionate love."

Mary gained insights regarding the deeper purpose behind events in her life, even uncomfortable and "miserable" ones, which now due to her greater understanding had a sense of beauty. She explained, "Then I was also shown the beauty, *the great changes and ripple effects* . . . and that beauty comes of all things," even seemingly hard and negative events in life.

Mary's sense of time also changed during the life review, in a way that she (and most people who've experienced this) struggled to explain.

Dr. Mary Neal explained how she, like Dr. Robert Montgomery and others, too had experienced a vast expansion of her consciousness. She was able to simultaneously experience thousands of memories and images, whilst also fully experiencing them individually and independently.

While trying to explain *the vast and immense expansion of her consciousness,* she said, "It's a very strange concept. It's not really something that I can explain. *It is almost as though every moment in time expanded into all of eternity and all of eternity was experienced in every single moment.*" Dr. Montgomery had conveyed the magnitude of his hyperconsciousness by analogizing his ordinary day-to-day sense of consciousness to the "Earth" and his state of vast hyperconsciousness at death to the "cosmos." Dr. Mary Neal instead used the term *eternity* to describe the sheer vastness of what she had experienced of her hyperconsciousness at death compared with ordinary, day-to-day life. She acknowledged that "I realize that it doesn't really make sense, but that's really the experience." She further explained, "I don't have the language to communicate the intensity of this experience."

There is something about being in the vast hyperconsciousness state that provides greater insights and knowledge, not only about the cause-and-effect relationships underlying events in people's own lives, but also other realities. While in that state, Mary felt that she "had this absolute, pure understanding of the order of the universe." She said, "I understood how it all works. I understood math, which isn't one of my strong suits. I understood the natural world. I understood the truth of how we are each tightly interconnected and how all the billions of us on this planet and the even more billions of animals and plants and the entire living world are completely intertwined, and how what one person does or says or thinks reverberates throughout the entirety of our universe."

Throughout the hundreds of testimonies that I had studied and examined from the mid-1990s onward, this was something that I and others had often heard people say. As we have already seen, they frequently described gaining tremendous insights and knowledge about things they had not known before in their day-to-day lives. This included even complex scientific concepts, such as mathematics and quantum physics, as Mary had described, too. At first when I used to hear such claims of sudden insights and knowledge about the underlying fabric of the universe and deep scientific realities, I would ignore them as I had no way to explain them. They just seemed like oddities. However, with time I started to notice a pattern emerge as more and more people made the same claim. They described experiencing a huge expansion of their consciousness and with it a tremendous expansion of their knowledge.

Interestingly, though, as Dr. Montgomery had explained, their knowledge and those insights would be snuffed out, almost like a valve had closed, and they could no longer recall them after they had recovered, even though they knew they had come to know so much more then.

Meanwhile, as Dr. Mary Neal was having this tremendous inner experience, she also remained cognizant of things happening to her by now "dead" body, which was still trapped in the boat underwater. She noticed the force of the strong current of water was gradually pushing it out of the boat. "Eventually my body came over the front deck of the boat. And as it did that, my hips came around the corner and that was fine, but as my knees came around the corner, I could feel them breaking, because they had to bend backwards on themselves [to allow her body to be released]." Mary explained, "I took a moment. I thought, *Wow, I'm an orthopedic surgeon. I know what bones breaking feels like, I know what it feels like when ligaments tear and I can feel my knees breaking and I can feel the ligaments tearing and I should be screaming, but I'm not, I feel great.* So, I could feel all of this, and I was still in the present, but [at the same time] I was also having this incredible spiritual experience and this life review that was truly like nothing I personally could have dreamed up. Again, I was still me. I was still my analytical self, and as I was experiencing this, I was simultaneously able to sort of look at this and think, *Oh, my gosh, well, this is amazing.*"

Mary, like all the others who have reported observing their body while going through this experience, had also perceived that her selfhood—her sense of self—was different from her body. "When my body eventually came over the deck of the boat, I could feel myself [as a] separate [entity] from my body, and I never know what word to use for that. For the 'me,' [referring to her selfhood] was clearly separate from my body."

As her body was eventually pushed out of the boat by the current, she could understand not only what was being said and done—but also the deep thoughts and feelings that her rescuers were having. She said, "I could watch the guys do CPR. I could hear them call to me, *Please come back and take a breath, Mary, please come back and take a breath.*" Mary felt deep compassion and love for her friends who were doing CPR and trying to save her. She found herself going back to her body and trying to breathe (but to no avail) because her friends were asking her to.

Mary maintained lucid thought processes with reasoning and memory formation, despite the fact that she had drowned and now her heart had stopped and she was receiving CPR. Like many others who had also all recounted this to me before, her experience felt more real and more vivid than anything she had experienced before. As Dr. Greyson explained, importantly this was happening "in the context of severely restricted oxygen to the brain. When you would not expect much of an experience at all, let alone hyperacute senses and rapid clear thinking."

Mary described the greater sense of reality and vividness of her consciousness compared with her ordinary day-to-day consciousness, which—as others too had often described—felt like a shadow of reality. *"Comparing the two, it was sort of like watching a black-and-white television versus, you know, a state-of-the-art color, a 4K IMAX movie. It's just very different."*

Even though people were still performing CPR on her body and trying to save her, she felt conscious and more alive than before. She explained, "I knew too much time had gone by to still be alive, but *I never had that experience of feeling alive and then dead. I never had the experience of feeling conscious and then unconscious. I had the experience of feeling alive and then more alive; conscious and then more conscious.* That was because this experience was absolutely more real, more intense, more vivid than anything I have ever experienced here on Earth, and I've experienced many things." She emphasized, "I loved my life on Earth, *but it is not the same.*"

As with the sense of separation, the life review occurs when people are unconscious and biologically either on the brink of death or have just traversed beyond the boundary of death. Think of it as "reliving a recording of life." That is because although people evaluate *every moment* of their own life and lucidly contemplate what was right and wrong about their actions from a moral and ethical perspective, it is as though all their actions, thoughts, and intentions toward others, and their inner states and emotions had been recorded in their own self (or consciousness) and somehow come out to the fore to be analyzed in detail with death. People experience their actions from the perspective of the other people involved: and the result is what one person termed *"the truth of the matter."*

Dr. Greyson shared the story of a resuscitated patient who remembered being seventeen years old and getting in a fight with a drunk: he started pounding the guy with his fists and he left the guy bloody on the street and then walked away. In his life review, he experienced it *from the perspective of his victim*, and he was able to count each one of the blows and feel his teeth being knocked in and the blood flowing, and also feel his rage from a third-person perspective.

During the review, there is a sense of being in other people's shoes and experiencing their perspective of events in a *real, active, and dynamic* manner. *This means experiencing the exact emotions and thoughts that others experienced in response to one's actions, or equally as a downstream consequence—the domino cascade effect—of one's actions.* For instance, if a person's spiteful words cause humiliation in someone else or their flaunting leads someone to feel inadequate, then in death, those exact feelings that had been induced in others are relived.

Since the most minute details of one's actions in life are reexperienced in this manner, one can only start to imagine what the experience of death might feel like for those who purposefully and intentionally commit major offences against other people—whether directly or indirectly through others.

In equal measure, if one's intentional actions, whether directly or indirectly, led others to experience intense feelings of joy, happiness, and well-being—say, due to blooming personal or professional growth, learning, success, and so on—then those same exhilarating emotions in others will be experienced. That is together with their downstream effects through the positive development and well-being of their loved ones, which will also be relived and experienced. *This is the domino effect*, which we will discuss further.

Rachel, a young British poet and mother of four children, lost a lot of blood giving birth. She felt her eyes getting heavier and heavier and tried to force herself to keep them open. As her medical condition deteriorated, she realized, "I cannot keep my eyes open," and fell into unconsciousness. She said, "That was the point at which I was up and out of my body. I remember being surrounded by several people . . . They put an oxygen mask over my face. I remember that very vividly."

Like Mary, Rachel noticed "a presence." She found herself also "shooting" up and away through what felt like an "enclosed space," a "tunnel" of sorts. This was when she started to relive and evaluate her whole life. She said it felt like something resembling a large "holographic, translucent, and invisible" screen appeared. It was "not solid or physical. It was [so] enormous, [that it] encompassed my whole vision." She continued to explain that this metaphorical screen "was playing my life from a very young age, right up until the moments that led to me leaving my body.

"[These were] what you might consider really important and significant events, but also what you might consider really mundane events. I was watching them all and then I was reliving them in that moment, as if I just traveled back in time and straight into that second." She confessed, "It's still perplexing me.

"It was like watching this movie, [while] I am also kind of jumping into these scenes and reliving them, feeling and thinking and experiencing exactly what I did in each moment. [Then,] at the same time feeling and thinking and experiencing in that moment, from the perspective of whoever is there with me."

This was "really mind blowing." She said, "I was watching happy memories and sad memories and painful memories. I was watching the things I did that were good and the things that I did that were not so good."

She felt the "presence" had stayed with her from the time of "leaving my body and traveling through the tunnel" onward, but it "wasn't judging me or condemning me. It was just observing [my actions] with me."

She said, "In that state, and with my new awareness and with a perspective that was vastly different to the one I had while in my body, I was judging and exploring each moment, and [evaluating] how did I respond in that situation? Was that a good way to have responded? Could I have done that differently? Was there a better, more loving [humane] approach to that situation?" She explained, "I judged myself, but not with the thought process that I would have had with my body. I went back to [each] moment and truly relived it.

"If you think about some mundane thing that happened when you were, say, six years old, [which] you've totally forgotten, because it was not important. However, then to go back to that moment and experience it again with a different perspective and awareness—this was a really pivotal part of my experience."

She explained that the review did not focus "just on the more profound experiences," it involved even seemingly mundane events that Rachel could not have normally recalled. She also emphasized, like Mary, that although "I now know that is what they call a life review, I did not know that for many years."

She continued, "When I was about eight years old, we had a pet guinea pig. One day, while I was holding this guinea pig, I stood up, and he suddenly bit me. I was shocked that he had hurt my finger, and so as a kid, I threw him onto the sofa right beside me and that was it. My finger was fine, and the guinea pig was fine, and I didn't think any more of that incident. Now, during this review, I was watching and reliving that moment, which in my life had felt irrelevant, and unimportant, [but I saw that] it was very important. I experienced *how the guinea pig had felt* when he had bitten me. He had become afraid [because she had suddenly stood up] and my throwing him on the sofa [afterwards] did nothing but create more fear in his being."

Rachel explained that she felt shame "even though nobody was shaming or condemning me" at how she had hurt the guinea pig, and it had changed how she lived her life. "That is where I learnt an awful lot about what I now consider is important to being human and the purpose and relationships that we share, which was very educational for me. I brought that knowledge back with me."

She mentioned that she rewatched and relived things that she "had suppressed or forgotten" and said that "was a shock to my system."

As Mary and Admiral Beaufort had also experienced, Rachel, too, came to learn the "cause and effect" relationships—*the reasons*—behind the events in her life, which she had not previously understood. She said, "I was not just looking at it [the events] from [an ordinary] human perspective," instead "[I was] looking at it with awareness of [the reasons] why that [event] may have happened." By that, she clarified, she meant "the purpose, and the lessons that can be taken from that experience, and the spiritual growth [that comes] from a whole number of different scenes that I relived in that portion of the experience."

She described some intimate, personal, difficult, and deeply hurtful details about what someone had done to her, which she had learned during her review experience. This hurt, but it helped her to gain insights about the *chain of cause and effect* that had led "other people to do wrong." *Without excusing other people's errors (as Mary had also said), she understood the reasons behind those mistakes and what lessons could be learned from them.*

She explained, for instance, "I had an argument, or disagreement, with somebody who's very close, and I was reliving that and then *reliving it from their perspective*, and in that perspective, I [find that I] have a greater understanding of their past and their history, their difficulties, their past trauma, and their development [background] and everything else that led to their choice. During the disagreement, I was understanding it from my perspective, but then also reliving it and understanding it from their perspective, *and then understanding why that was their perspective, and why it had to happen that way.*"

She understood the person's background and life events, the chain of events in their life that had led to their erroneous and hurtful actions. "Perhaps they didn't have a greater understanding of their choices, or how their behavior was going to affect somebody." Likewise, she understood more about her own behavior. "I would see where I had been a spitfire with my words while I had been cross and angry. I was then feeling myself living in the perspective of the other person, who had been at the receiving end of the brunt of my harsh words, as I was launching them at them, without any real thought. I was nineteen and I was hotheaded, fiery, and selfish. It was only through reliving those moments that I gained a really good understanding of how that behavior affects other people." She went on. "If I am spitting cruel words, let's say on my sister because we were in the middle of an argument, and then in the same breath, I was feeling exactly how that made her feel. I would be feeling sad, angry, hurt, ashamed, embarrassed, belittled, all of these things, just from one quick insult." She explained, "And then also gaining the understanding that she was going to carry that with her and that it would still be with her for years later. It's not until you feel that that you can appreciate the enormity of it."

She realized, "I'm going to say one horrible thing and I think nothing of it, but that stays with the individual. That was what really shocked me. It was to see that we are all carrying [some] trauma that we don't even know is from every interaction we share. We are insulting people or putting them down and not actually understanding how deep that goes. So that is going to stay with that person, and that's then going to change their whole perspective of themselves, because somebody else has picked up on something that they may not like about them.

"That stayed with me and I am glad it did, because I had a number of years with my father before he passed away. After that experience, I would sit and think to myself in my mind that *you must control your tongue*, because now I knew how that was making him feel and there was no need to respond like that." She confessed, "It's really hard to explain."

This experience transformed Rachel. It taught her why compassion, caring, and justice are important. It also taught her that "love is the root of everything and can heal everything. The whole purpose of life is to bring love to this plane."

Some people have told me that while there may be comfort in knowing that understanding comes from this experience, the prospect of feeling exactly how they had made others feel, together with the downstream consequences of their actions, is daunting and frankly frightening.

Interestingly, though, while people unanimously seem to describe feeling the same emotions they had inflicted on others, including the sense of humiliation and shock their actions cause, *their reaction is always positive and educational.* Under normal circumstances, having to relive one's painful actions and their impact on others might lead to a negative inner state and outlook about oneself. However, in death this instills a positive inner state and outlook that drives people to make a positive change in their behavior in order to improve as a human being. This is why Rachel and so many others have described their experience as having a transformative effect. In Rachel's case, it led her to change her behavior toward her father and she treated him with far greater compassion and understanding.

Of course, the overall conclusions drawn from people's testimonies here are based on what people like Rachel and Mary have described—those among us who are not purposefully trying to hurt others. As mentioned, one would imagine the experience would be much more painful and distressing for people who purposefully set out to hurt others.

Ultimately, the experience has a profound and positive effect on how people conduct themselves, no matter what their initial circumstances. They view it as an opportunity—a second chance—to make up for any prior errors.

Chapter 11

NEEDED TO DO MORE: NOT AS GOOD AS I THOUGHT

RACHEL FINCH'S AND DR. MARY NEAL'S TESTIMONIES, WHILE incredibly powerful and insightful, represent two out of many hundreds of testimonies with descriptions of the review of life that our team of researchers at New York University have studied.

Through our analyses, using the scientific method known as grounded theory, which I explained earlier, we have identified many more highly meaningful and purposeful themes within the context of the review that together provide a greater understanding of this intriguing and totally unique aspect of the recalled experience of death.

Based on our analyses, what Moody and others have called the life review can be broken down into at least twelve specific subthemes:

1. *Reviewing a recording of life*, in which nothing appears hidden—all thoughts, intentions, actions, inner states, and emotions throughout life appear to have been recorded and stored in a person's consciousness and are recognized to have mattered.

2. *The indescribable.* Being in the presence of a compassionate, loving, "perfect," and luminous being.
3. *Hierarchy of knowledge.* There is recognition of higher levels of knowledge and wisdom, which relate to having higher fields of perception to understand those truths.
4. *Whatever good and whatever harm I had done was to myself.* There is a realization that harming others ultimately reflects back on the person themselves and vice versa.
5. *Relive life events and reexperience each moment.* Specifically, people are able to reexperience themselves in all events in their life from the most mundane to the most extreme.
6. *Need to do better: not as good as I thought.* People come to judge their true worth as a human being based on the principles of ethics and morality. They recognize their purpose in life had been to improve themselves as a true human being.
7. *Being in others' shoes.* People examine their experiences in life from multiple perspectives, including those of the people they affected.
8. *Evaluation of actions and inactions in life.* Not only actions are evaluated but also times when people could have done something positive for others but did not.
9. *Domino effect*, regarding the impact and downstream consequences of their actions—both good and bad—on people's lives. They appreciate, as one person said, "how big an impact my seemingly small actions had on a large scale."
10. *A different value system.* Seemingly insignificant actions in ordinary life, such as giving someone a helping hand, are discovered to have been far more valuable than many of the things that people (and society) present as being important.
11. *A reason and cause and effect underlying all events.* People appreciate that everything that happened during their ordinary day-to-day life had a reason. Specifically, they recognize that the principle of cause and effect, which governs everything else in existence and is the cornerstone of science, also applies to events in their lives.

12. *Higher overall purpose to life.* They wish they had understood and appreciated their higher purpose throughout their lives so they could have lived their life more meaningfully and shaped their destiny better. As one person said, *"I saw that I alone am in charge of my destiny."*

The following cases break down the different components of *reviewing a recording of life*. These include features such as *feeling embarrassment and shame*, in which people recognize they could have lived a more moral and dignified life. One summed this up by saying, "I had done so little with my life! I had been selfish and cruel in so many ways! I was truly sorry I had done so little." Overall, there is a recognition that life had comprised *an education*, with the purpose of helping people gain knowledge about how to *improve as a human being.*

1. REVIEWING A RECORDING OF LIFE

The following cases highlight the overall sense of *reviewing a recording of life*. Marta was admitted to a hospital in Guadalajara in 1985 and had a severe, life-threatening allergic reaction to medication. She explained that she lost consciousness, and then *"I saw my entire life in great detail, and experienced feelings through it of satisfaction, shame, repentance."* Another person said, *"I saw and experienced every single detail of my present life up to [the moment of undergoing a cardiac arrest], like watching a movie, yet starring as the main character simultaneously . . . I was my own judge."*

Richard lost consciousness after an electrical malfunction in his heart caused it to stop in 1985. He said, "I started watching my whole life being reviewed in front of me. The emphasis seemed to be on good and bad events . . . "

Lou, a retired US Navy officer who survived a life-threatening car accident in 1963, explained that while unconscious, "in an instant, my entire life, starting from birth through the present, was flashing before my eyes . . . Frame after frame, some parts in freeze frame if only for a second, then on to the next. I felt as if I was being subjected to a test . . . "

Karen underwent a spinal fusion procedure in 1982. She was thirty-two years old with two small children, ages six and nine. The last thing she recalled was being given a general anesthetic and then losing consciousness. Later on, during her surgery, she experienced her consciousness as separate from her body and near the ceiling, with full awareness. She was observing her surgeon frantically trying to stop her from bleeding to death. Due to the stress, he was swearing at a nurse to fetch blood immediately. She said: "I was watch[ing] my life . . . I was able to reexperience myself in all these events in my life, but just as importantly, I was able to experience the impact of my actions and words on those other people with whom I had interacted . . . I felt so very sorry and sad about them [describing her poor actions] within my own heart. It all seemed to happen very quickly but had a tremendous impact on me."

Arshan added, "My life and all of its events started to play in my mind, but it was very clear, real, and alive. It was like a slideshow, *but I experienced all the feelings in these events again*. Everything was shown in chronological order. Although this whole life review only took minutes, it was pleasant and interesting to me. Once my life review stopped, my mind started to analyze my life and my actions. I felt that overall, I was relatively kind to people."

As Arshan and others had highlighted, the evaluation is based on being in other people's shoes and learning directly how their own actions had impacted others. Arshan learned he had been kind by experiencing it from other people's perspective. One other person described the sense of being in other people's shoes as follows: "[I was shown] everything good and bad that had taken place in my life. In your life review, you switch places with the people you hurt. You feel the emotional pain you've caused them. It was painful to watch how I had hurt people . . . "

2. THE INDESCRIBABLE

As already described, the evaluation typically takes place in the presence of a compassionate and benevolent entity—a "personified luminous being"—who is perceived as having an extremely high magnitude of power and a far higher field of perception and knowledge of truths, imbued with indescribable levels of kindness, compassion, and humility. Its magnitude leaves people in awe; people find it to be somewhat *indescribable*. This entity helps the person who is undergoing a review to draw appropriate lessons from the evaluation of their own actions and intentions. And it is this entity that people refer to when they talk about seeing and experiencing a "light" that was full of love. *It is not an abstract light. It is a personified luminous entity with enormous magnitude.*

In 1974, Anni was thirty years old when she suffered a life-threatening complication during a medical procedure that rendered her unconscious. She explained, "I was shown my life from birth to unconsciousness. I saw myself on the wrong side. I was not as good as I thought I was and was ashamed of myself. But the being of love didn't judge me. He just supported me and gave me love. I saw not only the actions I had done, but also the thoughts I had sent out. And the thoughts meant more than the actions. That surprised me. I hadn't thought it would be like that. It was scary. It's very good to do good deeds towards others, *but the feelings and thoughts you send to them count more*. For instance, it is bad to smile politely at someone while sending negative thoughts to them. *I found out that there was so much that I had to do. I had to improve as a human being*... I was also shown the good things I had done."

This represents one of many cases where people emphasized evaluating not only the quality of all their actions in life, *but also the quality of all the thoughts and intentions behind their actions. There is recognition of a tremendous value attributed to the intentions behind actions.*

3. HIERARCHY OF KNOWLEDGE

People sometimes recall meeting more than one luminous entity, and there is a *realization of the existence of a hierarchy of knowledge and understanding (reflecting differing fields of perception) of truths, even among the personified entities of light, all of which ultimately come from one unifying and originating source.*

Katherine, an American woman who described herself as agnostic when she suffered a life-threatening ectopic pregnancy in 2001, recalled there were "two beings who guided me . . . for a life review process. After the life review, I was taken before more beings, [who] seemed to be wiser [than the first two]. I communicated with them about my decisions during my life review and areas where I could improve. While it was a collaborative process, I had deep respect and reverence for these beings." She then described the extreme compassion and kindness—which many people refer to as "love" [or compassionate understanding]—emanating from them. She said, "I felt that they loved me completely and without any judgment." She said, "In psychology there's a term to describe this. [It is] called 'unconditional positive regard.' I felt completely sure that they had this feeling for me. This surety felt like a warm glow of light around me."

The hierarchy is gentle, and not oppressive, but there is a clear sense that there are layers of comprehension within this world. "I felt sure that the person made of love was much, much, much more superior to me. I understood they were beings in a much higher level." Tricia, a woman whose heart stopped during a routine operation, recounted, "The [beings] were trustworthy and there to help and comfort me, so I did not question their authority."

Our subjects consistently reported knowing they were lower in the hierarchy of knowledge within the beings' world. They recognized that some of the beings of light held wisdom and knowledge far beyond their own *understanding, even though they were in the state of vast expansion of their own consciousness.* All our research subjects reported being at peace with and accepting that their knowledge was highly limited compared with these beings'. Their experiences were characterized by deep humility and no sense of ego whatsoever. *They recognized that through effort, they could expand their own field of perception and hence understating too. This was part of their overall purpose in life, which they had recognized.* In fact, none of our interview subjects reported being angered, frustrated, or feeling any negative emotions about it. Instead, there was simple acceptance that they were much lower on the hierarchy, but that this is a place where beings, including themselves, can ascend, as this is a place of learning truths. "My own awareness of this new dimension seemed much more limited than their awareness. I was taken before more beings which seemed to be wiser than [others]." Leonard, an older man whose heart had stopped, said, "The [being] seemed to be in communication with a higher power. Every so often he paused to listen, or to speak upward. A decision had been made. [The beings] asked who I was. 'I am Leonard K,' I said. I felt incredibly small and overwhelmed when I spoke my name." *This further highlights the immense sense of humility that is experienced, which I just mentioned.*

Interestingly, some reported a sense that there were also other beings below them in the hierarchy. Mohammed said, "There were other [beings] there too; some with more light and possibilities than me and some with less." *This further highlights the existence of a hierarchy of knowledge.*

4. WHATEVER GOOD AND WHATEVER HARM I HAD DONE WAS TO MYSELF

One other person described the compassionate, luminous entity as follows: "[I had a] feeling that there is a presence that was following me all the time . . . he showed me . . . scenes from my own life. The scenes were in chronological order from the very beginning of my life . . . *I saw that whenever I had done something good to anyone or anything, that I had done it to myself. And whenever I had hurt someone, I had done it to myself.*"

Ron, who had suffered a life-threatening car accident in 1962, explained this point further. He said, "During the review, we revisit scenes from our lives and feel the actual pain or anguish, pleasure or love that we have inflicted upon others. We become the object of our actions . . . *The purpose of the review is not for punishment, but for growth through understanding the ramifications of our actions, thereby gaining increased compassion for others. The ultimate irony, however, is that every time we hurt someone else, we eventually hurt ourselves.*"

5. RELIVE LIFE EVENTS AND REEXPERIENCE EACH MOMENT

Bolette explained: "I met a glowingly beautiful, very loving being . . . His loving presence completely surrounded me and together we went through my life and all that I had experienced in a loving way, not in any judging way. It was observed, and all the feelings involved during life were examined . . . All situations were examined, and all the good was emphasized and shown. I could see it with him."

Sharon, a middle-aged American woman, recounted how in 2005 she had been hit by lightning while outdoors. She said the pain was excruciating and she lost consciousness as her heart stopped. During her experience, she says, "I felt a huge presence all around me just pouring love out onto me. I felt such joy and all I could do was stand there in awe at the beauty and the love that was all around me . . . then . . . I was given my life review. I was shown my life; *everything I had ever said and done was shown to me.*

Another person said: "[The benevolent personality] is aware of everyone and everything every minute of each day, that each act, word, intention is duly noted."

One person explained: "Not only was I viewing moments [of my life], I was feeling them happen again as if I were there." Another said: "[A being of light who] told me he was there to help me . . . started to show me my life like a movie . . . The first image I saw was something bad that I did. I could feel the pain that I caused because of my actions . . . This movie was showing, *second by second, my entire life*; everything I saw I could feel the results of it. Everything I did had a life of its own . . . He didn't show me just the bad things I did; he showed me the things I did out of love too [good things]." One other person explained: "I started to recall significant moments of my . . . life. It was like traveling to the past to be the main character of the story and an observer at the same time. I learned so many things about myself that I did not know." Another further clarified: "I was able to re-experience myself in all events in my life."

6. NEED TO DO BETTER: NOT AS GOOD AS I THOUGHT

Steve, an American man who had suffered a severe asthma attack that caused his heart to stop, said, "I found there was a being beside me. I could feel his presence. It was a comforting presence, a reassuring presence but was also a presence of magnitude and power. I felt things were all right. Then I began a review of my life . . . But at the same time, I was re-experiencing it from the other people's points of view and that was a stunner because you feel their pain, you feel the sting; you feel the hurt [caused to others by your actions]. It was a horrifying realization that I wasn't the person I thought I was."

Much like others, Steve had also experienced firsthand the same pain or discomfort and emotional distress that he had caused other people. He relived those moments, from his own perspective, but also from the perspectives of others, by experiencing exactly what they had experienced. Thus, he was able to judge and evaluate his own actions and behavior toward other people—by feeling their effects—without the need for an external judgment. It was all internal.

With parts of his life laid bare, he explained how he had been shocked to learn how badly he had sometimes treated others. He realized during his experience that, at times, he had been deceitful, had hurt people, and had outright lied, all of which he had justified to himself in his day-to-day life by rationalizing that people deserved what he had done. But experiencing other people's pain as they had experienced it made him question his own morals and humanity and made him conclude that the way he had lived his life had been a failure.

He explained, "I re-experienced these events [referring to his actions in life] from my own point of view. I wasn't just watching the events; I was actually reliving them again, while at the same time, I was also re-experiencing the actions from other people's points of view. *I was them.* I was reliving the experience from their point of view, and at the same time, and I don't know how this works, *I was also experiencing it from a higher reality, the truth of the matter.* So what I saw was my own lies and my own self-deception to myself, which I had used to convince me that doing certain things was okay because people had deserved it. Then I was now experiencing the emotional impact it had on other people. I felt their pain. I felt the shock on them . . . So the net result was that I felt like I was a failure as a person, and I wasn't the person I had thought I was. It was humiliating. I felt really dreadful, and it was completely humbling."

He emphasized, "*The judgment came all from myself. It was not from an outside source, but then this being that was with me was also sending me comforting messages.*" He explained that the entire experience made him feel as though he was given a second chance to live a more meaningful life. He felt he had a chance to change things so "*next time I get back to the life review it wouldn't be the same, or at least they would say he tried,*" and then, half-jokingly, he said, "*I would receive a higher grade!*"

7. BEING IN OTHERS' SHOES

Justin is a British man who survived a life-threatening accident in 2004 that had left him in a deep coma on life support for many days. He explained, "Many events in my life I experienced, but not from how I remembered it, but [instead] from the point of view [of] how the people, animals, environment experienced it around me. I felt it as my own." This case, like Rachel's, highlights how *people experience being in the shoes of not only other humans, but also animals and all other living beings whom they may have impacted. Everything is laid out and reexperienced.* He continued, "The times I had made others happy, and sad, *I felt it all as they did.* It was very apparent that every single thought, word, and action affects everything . . . In the life review we judge ourselves; no one else does, the light [luminous personified entity] did not." He then confirmed what Steve had explained when he said, "*But with no ego left and no lies, we can't hide from what we have done* and feel remorse and shame, especially in the presence of this love and light. *Some of the things in life we think of as important don't seem to be so important there. But some of the insignificant things from the material human perspective are very important.*"

One other person explained: "I went through a life review . . . It was all about my relationships with others in this review. During this, I felt what they felt in my relationship with them. I felt their love or their pain or their hurt, by things I had done or said to them. Their hurt or pain made me cringe and I found myself thinking, 'Oooh, I could have done better there.'" Another person said, "I was shown my life in review from the perspective . . . [of] the truth. I was shown every time I had been selfish, choosing for my own interests. *I was shown every time I had been divisive or manipulative for selfish gains. I then felt that pain several folds over.*"

As Mary, Rachel, Karen, Steve, and many others had recounted, the evaluation occurs from three standpoints: (1) their own perspective, (2) other people's (or other beings') perspectives, and (3) the perspective of the *truth* of the matter—how the issue had really been. This means independent of how people had perceived the specific events in their life, they learn the real truth. This includes detailed events that the person would ordinarily not be expected to recall.

People also evaluate opportunities—no matter how big or small—when they had acted in a higher and more dignified and humanistic manner toward others, as well as opportunities—*again no matter how big or small—when they had failed to act in a higher and more dignified or humanistic manner. The review also includes what people should have done but had failed to do.* Ultimately, through the review, people typically recognize they had *not been as good as they thought.*

8. EVALUATION OF ACTIONS AND INACTIONS IN LIFE

Katherine explained, "I reviewed my life like a movie except that I could pause it and zoom into different important times during my life. I could examine these times from multiple perspectives, such as the people they affected." In her review, she learned that "it wasn't so much a decision of doing the 'wrong' thing in situations, or making unwise choices, *but that the times of greatest challenge for me were times in which I could have acted but chose inaction.*" This was an important point. Others have commented that they too had analyzed what they had done—their "actions"—and also what they had failed to do—their "inactions." For Katherine, "it was concluded, that when I returned to Earth, I must choose action and use my experiences and feelings to guide these actions."

Heather, a British mother, recounted a similar experience after undergoing a life-threatening surgery. She said, "I suddenly found myself standing beside myself looking at a cord which connected me to my body and thinking how thin and wispy it was. Someone was beside me. I was made to feel secure and encouraged to trust my companion, who suggested that the cord was insignificant and that I should not concern myself with its fragility. I was guided towards the light. This was a sort of void, in which I found myself with the ability to fly, or should I say I had no weight—a very strange experience. Throughout the journey I kept looking back to ensure my companion was with me but somehow towards the end of the journey I found myself just content to move on and reach the end.

"Reaching the light [described as a personified luminous entity], I was met by other beings of light [other luminous entities] and very gently encouraged to move on towards a life review. In this experience my actions were not judged by others, I judged myself. My presence could see into my mind and *there was no way I could hide any thoughts*. Gently, I was encouraged to understand how my mistakes hurt others by experiencing what others had felt as a result of my actions. I was confused, as it all seemed so strange. The word 'death' was never mentioned, yet somehow, I came to understand that I was in that place where the newly dead move on to."

She told me that during the process of judgment, she felt uncomfortable and remorseful about the opportunities she had failed to make use of in her life. She described these as situations where she could have had a positive effect on others but did not follow through. She also told me that she now hated to do any harm to others, as she had felt the pain she had caused others. She now felt that the most important thing in life was to take the opportunities to be of assistance to others, even if these were sometimes difficult options. This highlights the positive transformation that many people undergo following their review, particularly after encountering the luminous personified being who helped and guided them through their review.

9. DOMINO EFFECT

During the review phase of the experience, people also recognize and evaluate the downstream consequences of their actions on other people's lives. One man explained, "*I caught glimpses of my life and felt pride, love, joy, and sadness, all pouring into me.*" He said, "*I was shown the consequences of my life, thousands of people that I'd interacted with and felt what they felt about me, saw their life and how I had impacted them. Next, I saw the consequences of my life and the influence of my actions."*

As alluded to before, this case also highlights an important point regarding the magnitude of the review and, as this person stated, experiencing "*thousands*" of emotional states. It can be deduced that for those who knowingly and *intentionally* cause pain and suffering to others for their own perceived gain—whether to humans or animals—their experience in death is likely to be infinitely more painful than what we have come across so far. They will potentially experience thousands or millions of painful emotional states, which may include experiencing the emotional states of the people who have been hurt by their actions. *So one bad action could lead to an exponentially higher number of painful emotional states, depending on how many people it directly and indirectly impacted.*

Of course, as already mentioned, this works both ways. Equally, if someone had purposefully and intentionally been of benefit to others, they will likely experience thousands or millions of very positive emotional states. *This highlights the magnitude of the purposeful and meaningful review.*

Others also described that they learned how they should have acted differently. One survivor explained, "There was a life review where it was like a re-living of certain moments in my life up to this point. I felt with complete clarity how I felt and how the other person felt through my actions, my words, and my thoughts. *These were times when I probably should have acted differently, used better judgment, not gotten caught up in emotion. This was a very humbling experience.*"

In the review people evaluate every single event—*literally many millions of individual, previously forgotten moments*—in life. This includes the little deeds, forgotten thoughts, and seemingly inconsequential moments. One person explained: "*I could see how the last moment of my life was a result of everything that had ever happened to me, before. There was complete acceptance, even of those moments that I remembered as less pleasant.*"

By evaluating and reexperiencing every moment of life from their own perspective, the perspective of other people, and that of the truth of the matter, people come to fairly reevaluate their own life. They judge their own true worth as a human being—in essence their moral and ethical stock and their humanity toward others—in an impartial manner.

As we have seen, this evaluation and review encompass the entirety of a person's life—every minute detail—rather than only focusing on major or important events that people ordinarily remember, such as the birth of a child, a graduation, a wedding, a job promotion, or only specific bitter or happy memories. This is a remarkable and inexplicable feature of the experience. That is because ordinarily we can at best recall a tiny fraction of our lives, no matter how hard we try during day-to-day circumstances. Yet in death, while the brain is severely deprived of oxygen and nutrients, there is not only a paradoxical sense of a vast expansion of our consciousness—as Dr. Montgomery had said, like the size of our planet relative to the rest of the cosmos—but with it also every single detailed memory from a person's life, the vast majority of which had ordinarily been totally forgotten. These are then evaluated and analyzed by the individual again. *However, at that stage nothing is left out: neither actions, nor thoughts, nor emotional states.*

10: A DIFFERENT VALUE SYSTEM

When people evaluate their life during their experience of death, their value system is different from what it is under ordinary life. In life, social parameters of importance and value are those that largely focus on enhancing a person's rank, status, comfort, beauty, wealth, ego, and so on. In death, people evaluate things based on a different value system altogether and give importance and prominence to actions and intentions based on human dignity and their moral quality. Specifically, actions and thoughts focused on genuine altruism, compassion, kindness, selflessness, and so on.

This has at least two ramifications. First, it suggests that selfless actions, which are often not given prominence or importance—especially in modern society—may be very valuable in death, whereas much of what is ordinarily valued by people is ultimately not considered truly important or valuable. This realization may also explain why people come back after their recalled experience of death with a different value system, focused on altruism and selflessness. Second, it suggests that people's recalled experiences are unlikely to be imaginary—as any imaginary experience would reflect a value system that people had been accustomed to in life.

Overall, it is also very hard to explain why people would go through a moral reevaluation at death and why in death their value system would suddenly become so different and totally opposite to what they had been accustomed in life. Not to mention, why the review would be this universal and reported similarly by everyone, irrespective of their background. In death, everyone seems to evaluate their life based on a universal set of ethical and moral principles.

In life, people find ways to justify their incorrect actions toward others. They may even create their own moral standards and convince themselves that the person who had been on the receiving end of their actions had somehow deserved it. Yet independent of people's beliefs and whatever moral standards they may or may not have held, somehow, at death, everything is laid bare and there is no room for self-deceit.

There are also many other nuances to this point. For instance, in all our detailed studies through the years, we have not come across anyone who has reported not agreeing with or wanting to argue or disagree with their moral evaluation. People don't report errors in this evaluation. This is despite the fact that when we are called out on our bad behavior in ordinary life, our first instinct, even from a young age, is to try to justify our actions.

Interestingly, during the recalled experience of death, shifting of blame is not reported. Instead, people accept who they had really been—no matter how uncomfortable, shameful, humiliating, and embarrassing that might be—without self-justification or self-deceit. In the words of one of our subjects, there is "*no ego left,*" and with "*no lies, we can't hide from what we have done and feel remorse and shame.*"

11. A REASON AND CAUSE AND EFFECT UNDERLYING ALL EVENTS

As Dr. Mary Neal and Rachel explained, there is also greater understanding and insight regarding other people's weaknesses and misdeeds toward us, which comes from understanding the "*cause and effect*" relationships that had led other individuals to do wrong. Although understanding those relationships may not excuse someone's negative actions, gaining knowledge and understanding about the circumstances that led them to commit their erroneous actions leads to compassion. As Dr. Neal explained, her otherwise "destructive" and negative emotions transformed to positive ones. *It also leads to an understanding of how those situations could have been used to better our own humanity. As a result, there was a beauty to it all*, as Mary Neal and others had also learned.

In my conversations with Dr. Greyson, he also acknowledged this based on his studies. He said that he, too, had encountered other similar accounts. Someone who had recalled being mistreated by her mother as a child had experienced those events "from the mother's perspective in her life review." This had led to an "understanding for the first time, [as to] why her mother felt compelled to do this, because of problems in her own childhood, and that made her totally feel differently towards her mother after that experience."

12. HIGHER OVERALL PURPOSE TO LIFE

Ultimately, during the review people come to realize there had been a higher overall purpose to their lives, which revolves around developing higher human ethical and moral qualities, summarized in the words of one person who said: "*I [learnt I] had to improve as a human being.*"

Ordinarily in society, we give more weight to people's actions than their private thoughts and intentions. *Yet as we have seen, for the person going through death, even their private thoughts and emotional states seem to come to the fore and are relived and analyzed.* This is in part why we concluded that a recording of all our thoughts and intentions, and also inner states, feelings, and emotions, is stored in our own self or consciousness during life and is somehow revealed at death through the process of disinhibition. It also suggests that, as some psychologists have proposed, the entirety of human consciousness may be much more enormous than what we can access in day-to-day life. Perhaps human consciousness is like a huge iceberg that is mostly inaccessible and hidden during life, yet inexplicably comes to the fore in its entirety at death.

This is why the recalled experience of death is unique and inexplicable and seems to emerge only in people who are unconscious and wading into the grey zone between life and death. Nonetheless, this hasn't stopped people from trying to explain it away as some sort of imaginary trick of a dying brain, and to dismiss and trivialize the significance of the experience.

Chapter 12

GOING BACK "HOME" AND THE FINAL RETURN TO LIFE

I REALIZE THAT YOU, AS A READER, HAVE NOW READ MANY INCREDIBLE yet consistent testimonies. These reports can be hard to reconcile with what we know of our world today, yet it is difficult to deny millions of people's testimonies; especially when these people are strangers to each other, from all over the world, yet they describe the same consistent features. History has shown us how dangerous it is for anyone to ignore the power of human testimonies. However, as powerful as these testimonies are, it may still be difficult for some to appreciate and accept them. This in large part explains why some people have been dismissive of the recalled experience of death.

Ultimately, I recognize that whether people are accepting or not depends in large part on how they come to view these testimonies through the lens of their own prior beliefs. As Dr. Chawla has also pointed out, how any new information is interpreted often depends on people's own starting beliefs, but this topic seems to be a particularly fitting example of this.

Throughout my life I have recognized the need to approach these testimonies, and the overall concept of the recalled experience of death, with the rational, logical mind of a scientist. I also recognized the shortcomings and irrationality of a dismissive approach to new phenomena when they are inexplicable or cannot be made to fit into scientific models at a given time.

So, how can we possibly explain some of these incredibly profound features of the grey zone, and remain within the world of a rational mind?

First, I think it is important to take a good look at ourselves and come to better appreciate the limitations of the tools and apparatuses that we all have at our disposal when it comes to trying to evaluate and understand reality in a rational manner. We all recognize how incredible the human brain is—it's a marvel—yet few of us recognize how inherently limited it is in the evaluation of new phenomena. Our brain likes predictability, pattern, and repetition. So, when it is confronted with a situation that is unprecedented, and beyond its processing power, it can sometimes struggle to respond. Our brain is superb at interpreting events in the three-dimensional world that we live in. Simply dancing, riding a bike, or even talking on the phone requires the processing power of a very powerful computer that is capable of handling huge amounts of data simultaneously. We even learned how the human brain optimizes its function by inhibiting the functions it doesn't need in the moment to gain the most meaning out of any event that occurs. It is truly a wonder.

However, our brain also has surprising limitations that most people may not be aware of. Look at any of Escher's optical illusions, or any optical illusion for that matter, and you will see how your brain gets confused when lines in an image are configured in a particular way. This shows that our brain sometimes cannot even interpret and make meaning out of simple two-dimensional or three-dimensional data as reliably as we may think it can do. Now imagine this same brain trying to understand and interpret a world vastly different from the one we live in or that we are familiar with. It is naturally going to be even more limited.

Not only is the brain limited, so are the sensory systems that we rely on to gather information about the world around us. For example, while the human eye is capable of detecting electromagnetic waves in only a very specific and narrow range (i.e., light waves only), a bee's eye or a snake's eye can detect wavelengths that lie outside the range of the human eye. Dogs have a far more sensitive sense of smell than humans, and so on. Thus, animals may detect certain realities that lie beyond the capabilities of humans. Anything that exists and emits electromagnetic (EM) waves beyond the very narrow visible light range, which our eyes can detect, will not ordinarily be perceived and knowable to humans except with the aid of technology that can detect the specific EM radiation outside this range.

Most people may not be aware of this, but physicists think that many other dimensions exist in the universe beyond our three-dimensional world. One of the best recognized scientists in this field is Dr. Lisa Randall, a Harvard physicist who, in her book *Warped Passages: Unraveling the Mysteries of the Universe*, discusses this in detail. Mathematical calculations have shown that within our universe there are at least ten dimensions in existence.

What do these new dimensions look like? What would it be like to live in a six-dimensional world? We can't even fathom such a thing. Why? Because our brain is limited to evaluate data in three dimensions only. We can't even draw a fifth or sixth dimension, and intuitively it makes no sense to think of anything beyond three dimensions. Based on these discoveries in physics, we have to recognize we are living in a small three-dimensional enclave in a larger, *multidimensional* universe. Our brain can perceive only what makes sense and is relevant to our life in our very limited three-dimensional enclave, but not beyond. Yet realities exist beyond this limited three-dimensional world. This is why we humans can't understand all reality that lies beyond us, even though we might think we can. I will explain what Einstein thought of these limitations in a later chapter, but for now let me just say that he absolutely recognized this.

People with a recalled experience of death describe experiencing a reality that is multidimensional and different from our three-dimensional representations. As such, it lies beyond the capabilities and limitations of our three-dimensional brain to fathom. However, in the grey zone of death, when human consciousness is suddenly liberated from the limitations of the brain, people find their consciousness is vast and able to appreciate other dimensions of reality. Why? Because it is not forced to operate within the narrower three-dimensional confines of the brain as it is in life. After they are revived back to life again, and their consciousness collapses back within the three-dimensional body and brain, as Dr. Montgomery and so many others have alluded to, the same brain restrictions apply, and cannot easily explain what they recall of their experience, or much of the insights they had gained.

In her book, Dr. Lisa Randall refers to a novel originally published in 1884 by the British writer and scholar Edwin A. Abbott to illustrate the difficulties we face when it comes to perceiving multidimensional reality. This novel, *Flatland: A Romance of Many Dimensions*, talks about a fictitious two-dimensional universe—where two-dimensional beings (of various geometric shapes) live. *Flatlanders* are beings who have lived their whole lives in two dimensions. They are shocked by any suggestion that there could possibly be three dimensions because they have never perceived such a thing. Everyone in Flatland thinks the universe cannot possibly be more than two dimensions. It is as obvious to them as it is to people in our world who believe that there are three dimensions as opposed to four or more in the universe.

The book's main character is called A. Square. He gains access to a new three-dimensional world called Spaceland. There he learns to appreciate three-dimensional things such as a sphere. But while in Flatland, he finds he cannot perceive a sphere again, because he is confined to the two-dimensional world again. Instead of a sphere he sees a series of flat two-dimensional discs that increase and then decrease in size, which are slices of the sphere as it passes through his more limited two-dimensional world. It is not until he leaves Flatland and enters Spaceland that he can truly appreciate a sphere for what it really is.

Like the Flatlanders, people with a recalled experience of death struggle to describe what they have witnessed: it is beyond the brain's ability to process and describe. This ineffable nature of the experience may in part explain why people are sometimes so unwilling to discuss what happened to them. If you can't find the words to explain the experience to a nonjudgmental, scientific listener, how do you describe it to someone inclined to mock or dismiss your experience? Ultimately, the recalled experience of death is almost impossible to put into words. Many of our subjects shared some version of: "There are no words to describe this experience" or "Words can't convey the feelings" or "I can't describe in detail because I cannot describe in words." "The euphoria was indescribable." "We have no words in our language for it." "My descriptions are as best as words can be used to explain it." "My limited human vocabulary cannot adequately describe [the being]." "I must try to explain what cannot be put into words." "The senses are heightened and don't really work the way they do here."

Our human languages, like our brain, were designed to describe familiar phenomena in a three-dimensional world. Our languages have developed to describe events and emotions in our world, and they can't expand to explain phenomena beyond it. In most languages, the words available to describe internal states are typically limited to a certain number of specific terms (e.g., *happiness*, *sadness*, *euphoria*, *light*, *darkness*, and *color*) with much overlap. So, if you think about it, although experiencing a bright light may occur with different events, distinguishing between these different experiences—as well as their effects and their impact on the individual through language alone—is very difficult. We need context to make sense of language. For example, experiencing a light may occur when someone looks at a light source, but this can also occur after developing an eye condition (as we shall see in chapter 16), a disorder of the parts of the brain that are involved with vision, the impact of hallucinogenic drugs (as we shall see in chapter 17), a deep transcendental religious and mystical experience, or, of course, in relation to death. While these are all different experiences, due to the limits of language, they may all be described using one common term: *light*.

However, even our languages are not all the same when it comes to how many words they have to express inner human states and emotions. Some languages are more expressive: I recently learned that Arabic, for example, has vastly more words than English. While scholars suggest the English language has around 600,000 words, Arabic is thought to have around 12.3 million words. Some scholars who have read the poetry of Rumi in the original Persian language have often told me that the inner states of love that he describes cannot be readily conveyed in an English translation. That is because there is a vast array of words and terminology to convey emotions that exist in some other languages, like Persian and Arabic, that we don't have in English. Language is not equal, even in our three-dimensional world. As a result, our ability to express ourselves is limited in our three-dimensional world, too, let alone when trying to convey something that is perceived to have been from a multidimensional state.

The recalled experience of death defies attempts to describe it. Just as A. Square could not describe a sphere in the two-dimensional world, so, too, individuals who return from the far shores of the grey zone find that an experience that felt like "home" while they were there is impossible to express with words. The language simply isn't there. That is why one of the other themes that we and others going back to Dr. Moody have consistently identified is that the recalled experience of death is simply ineffable.

There is also confusion as to how we interpret the words that are used. Most of the things we experience in the world are NOT ineffable: love is not ineffable. When a word like *ineffable* is used to describe a relatively prosaic event, it is stripped of meaning. Survivors often rely on the word *love* to express what they experienced, but they do not mean "love" in the way we use it during our lives. We have all had texts from distant associates or job applicants who would "love" to connect. Likewise, we have experienced romantic love. We "love" foods or movies or other pastimes. Love in the grey zone is a completely different thing, beyond any ordinary form of love that is experienced, vague preference, or shallow connection. As we have seen, during the recalled experience of death, the love is higher and far more intense than any type of love experienced in life and filled with compassionate understanding. The closest type of love to this, according to the survivors, is a mother's love, unconditional and full of understanding and care. The issue of language also impacts the experience of light, which people feel they must refer to using the term *love* or other similar words. This is ultimately why Bob Montgomery and others simply describe the overall recalled experience of death as being beyond earthly emotions. You can see why people feel "stuck" when trying to describe it.

Another example that illustrates the limitations of our understanding of reality and the social meanings we give to things comes from Plato. Years ago, when I had started my research, I came across a really good description of this in Jostein Gaarder's bestselling book, *Sophie's World.* In it Gaarder described Plato's famous allegory of the cave:

> Imagine some people living in an underground cave. They sit with their backs to the mouth of the cave with their hands and feet bound in such a way that they can only look at the back wall of the cave. Behind them is a high wall, and behind that wall pass human-like creatures, holding up various figures above the top of the wall. Because there is a fire behind these figures, they cast flickering shadows on the back wall of the cave. So, the only thing the cave dwellers can see is this shadow play. They have been sitting in this position since they were born, so they think these shadows are all there are.
>
> Imagine now that one of the cave dwellers manages to free himself from his bonds. The first thing he asks himself is where all these shadows on the cave wall come from. What do you think happens when he turns around and sees the figures being held up above the wall? To begin with he is dazzled by the sharp sunlight. He is also dazzled by the clarity of the figures because until now he has only seen their shadow. If he manages to climb over the wall and get past the fire into the world outside, he will be even more dazzled. But after rubbing his eyes he will be struck by the beauty of everything. For the first time he will see colours and clear shapes. He will see the real animals and flowers that the cave shadows were only poor reflections of. But even now he will ask himself where all the animals and flowers come from. Then he will see the sun in the sky and realise that this is what gives life to these flowers and animals, just as the fire made the shadows visible.
>
> The joyful cave dweller could now have gone skipping away into the countryside, delighting in his new-found freedom. But instead, he thinks of all the others who are still down in the cave. He goes back. Once there, he tries to convince the cave dwellers that the shadows on the cave wall are but flickering reflections of "real" things. But they don't believe him. They point to the cave wall and say that what they see is all there is. Finally, they kill him.

Plato understood the limitations that we all have of comprehending and appreciating things that may lie outside of our limited understanding. Since the other members of the cave had never believed any other form of reality could exist, they wouldn't even consider it. They all had the potential to see what one of them had seen, but instead they attacked him. Based on what they believed of reality, they thought he had turned into a threatening deviant.

If a limitation is inherent because of the body's physical limitations or limitations of thinking due to an ingrained habit of thinking, how can we get around it? It is very difficult to persuade people to think about reality in a different way if they are unable to perceive it or they have a fixed opinion. How does one explain color to someone who has been blind from birth, or music to someone who has been deaf from birth, or taste to someone who has never tasted?

Perhaps we can explain the concept of sound as wavelength, or the taste of an apple as being sweet and sour. This may allow an intellectual understanding to be reached, but this can never compare with the level of comprehension that comes with seeing a beautiful painting or hearing beautiful music or tasting an apple.

Throughout our testimonies, I have been grateful to people who are willing to step outside the cave and return to describe what they saw there. Along with what we have reviewed so far, they describe other specific experiences and observations that relate to being back home, followed by their return to life again. As we are about to see, this travel back home component of the recalled experience of death can be divided into at least twelve distinct components, or subthemes, some of which have continued from the separation:

1. *Traveling back "home."*
2. *Traversing through a metaphorical tunnel.*
3. *Traveling at very rapid speed toward a luminous place.*
4. *Returning "home": a place of perfection I had known.*
5. *Time: nonlinear and not as it would seem.*
6. *A glance at my past and prior past.*

7. *Communication: thought is everything.*
8. *A source, an origin.*
9. *Feels more real: ordinary life is like a dream.*
10. *Learning the importance of making amends for infringements.*
11. *Acquiring knowledge of the future.*
12. *Suicide: not an end and not an option.*

1. TRAVELING BACK "HOME"

People who have a recalled experience of death sense that they are being pulled away by a very powerful and irresistible force, one that is imbued with kindness and goodness and pervades through them.

They describe a realization of heading back "home" to a familiar place. The review and reappraisal of life is typically recalled to have taken place during this stage of the experience. It happens either as they travel from their body and the vicinity of the room through a metaphorical tunnel of sorts at tremendously high speeds, or when they reach their destination, all while the senses continue to feel immensely heightened and much sharper than usual.

In 1981, Keith was a teenager and had taken some illicit drugs, which had led to an adverse reaction in his body and eventually caused him to suffer a cardiac arrest. He explained, "I went into a . . . tunnel. I felt no pain, only complete calmness. My whole life passed by me while I was traveling down this tunnel." An American woman, Kerry, suffered a life-threatening allergic reaction (anaphylaxis), and her medical condition deteriorated rapidly. She went into a state of medical shock, and her blood pressure dropped to dangerously low levels before her heart stopped. During the ensuing state of seeming unconsciousness, she recounted: "I found myself pulled up through the light at an accelerated rate of speed. It was like I was being sucked through the air by a powerful, yet gentle and loving force. I wasn't afraid, though. All I could feel was love that was so powerful that I knew I was going home."

Another American woman, Sharon, explained, "I went up through the ceiling, past the roof, into the sky . . . I felt so warm, safe, protected, and deeply loved."

Ron said, "As I hovered, I felt a wonderful force beckoning from above. I was going home. All I had to do was will it and follow the force, or, rather, let it draw me up."

2. TRAVERSING THROUGH A METAPHORICAL TUNNEL

While attending a car show in 2005, Rob, a forty-seven-year-old American man, went into cardiac arrest. He was revived through CPR and explained, "I remember entering a tunnel... I was slowly moving through the bright tunnel... Then there was the desire to be home... I was on the way home to where I belong and where I came from."

Someone else from England said, "I found myself well above the operating theater (room), where I should have been on a floor above that room or outside looking on a roof, but I wasn't. Instead, I was floating in the entrance to a tunnel." One other person explained, "I died and floated slowly and calmly out of my body, as though it was the most natural thing in the world. After floating under the ceiling, I left the room, leaving behind my body, the hospital ward, and the hospital where I lied down with my newborn son." Someone else explained, "I continued to float up and a tunnel appeared. There was a beautiful tunnel with a bright light at the end of it."

Graca, a Portuguese woman, suffered an anaphylactic allergic reaction that caused her heart to stop in 1982. However, fearing ridicule, it took her fifteen years to confide in her husband. She described "a sentiment similar to one that we have when after a long absence and we come back home. I felt I was going back home. I was at peace and as happy as I never had been before." Someone else explained, "I felt very relaxed and peaceful. Everything was black and I had the feeling of being in a tunnel. Way off, in the distance at the end of the tunnel, I could see a bright light."

3. TRAVELING AT VERY RAPID SPEED TOWARD A LUMINOUS PLACE

The sense of traveling through the metaphorical tunnel and toward a place of light is perceived as being exceedingly fast. People often say it feels faster than the speed of light. Heather, an American woman, had suffered a life-threatening heart malfunction that caused her heart to stop. She said she found herself "falling backward through a . . . tunnel at what seemed like thousands of miles per hour . . . It was a tunnel with so many lights."

When describing the speed, Lisa said, "Then I was swept away by some unknown force, and started to move at an enormous speed, which felt a lot faster than the speed of light." Another person explained: "I remember rushing into a dark, cave-like area where I continued at high speed for some time before I became aware of a small bright light in the distance in the direction that I was headed." Someone else said: "I briefly hovered over my newborn baby . . . then I was traveling . . . It felt like I was shooting through a tunnel, but I couldn't see any sides to it . . . I was tumbling, forward/upward at an unfathomable speed."

William, who had been a fifteen-year-old American teenager when he had a motorbike accident in 1978, explained, "I went upward . . . and accelerated faster and faster . . . it appeared as though we were going through a tunnel of light." Telesa described how "[I was] flowing with the light. The light carried me off . . . into a ball of light so bright that I could see it long before I actually reached it." Anni said, "I went through a tunnel. At the end of the tunnel there was a light so indescribably strong . . . "

Rick, a keen hunter, fell eighty feet from a tree stand into a river. As he was falling, his thoughts turned in slow motion toward his family. Then there was nothing as he hit the ground and lost consciousness. At this point, he said, he suddenly experienced leaving his body, before "traveling at a high rate of speed upwards . . . Faster and faster, [as] the speed was increasing!" He continued, "I saw a faint light growing brighter and brighter in the distance up ahead."

4. RETURNING "HOME": A PLACE OF PERFECTION I HAD KNOWN

People describe arriving at a place that is imbued with a vast sense of warmth, kindness, and goodness, which ultimately feels like "perfection." This sense of sheer perfection and the indescribable love and "light" that emanate from everywhere leaves people in a state of bliss that is so powerful that they typically do not want to go back to their ordinary lives—even if they should have had a reason to want to go back, such as a young family. One of the other reasons that they do not want to return to their lives is that they come to perceive that their ordinary day-to-day life had been a mere shadow of this new and far more vivid and blissful reality.

Stefania explained, "It was a place that had nothing to do with any kind of earthly or material experience. I was floating in a dimension which was enveloped in a soft celestial light, very rarefied. It was a light full of quiet and peace. It expressed unconditional bliss without end . . . Nothing could ever be as perfect and soft like this heavenly condition of absolute and far-away bliss in which I was floating." This highlights how people use the term *light* with positive adjectives—such as *warm*, *loving*, *compassionate*, and so on—rather than as an abstract, characterless light, or an entity with negative qualities, to also describe the place they travel to.

Joan, an American woman, said, "I simply had this profound experience of love, oneness, and freedom. I went from being in my body to being in a place of absolute Love." Another person said, "I had no pain, wants, or needs of any kind. I had a sense of being home . . . There was an understanding of complete peace, happiness, and contentment without need or wants . . . I felt this was home, where I came from."

Bobbi, an American woman who said she had a recalled experience of death in 1976, when she suffered surgery-related complications, explained, "You feel that you are finally home, where you belong. Now, I never had a memory that this place was home. It was more of an instinctive and intuitive knowing that this is where I belonged. Doubts don't exist there. They are not possible. Everything is perfectly clear knowledge."

Rachel, the British mother of four whose testimony was described in detail, explained, "It was immediate peace. Absolute, whole peace all throughout me. There was no pain, there was no fear, there was no shame. I felt completely accepted. Totally whole and loved. Loved beyond comprehension. Loved in my entirety. Loved with a Love I have not felt here. Loved with the purest love [and understanding] there can be." Jeff also explained,

"The light was love and understanding. It was outside of me, through me, and in me. It was home. I've never felt a love like this since, though there have been very brief moments of kindness and acceptance that I just live for."

Rob, the forty-seven-year-old man who had suffered a heart attack and cardiac arrest while attending a car show in 2005, explained, "Try to remember the first time you saw your child or met your significant other. It is that feeling of first-time love that is so positive and so powerful. Now take that feeling and multiply it thousands of times over. It is a love that is unimaginable."

Ron, who had suffered his cardiac arrest in 1962, explained, "This Love was so powerful, so extremely fulfilling—everything else was immaterial. This all-mighty power of Love goes well beyond our egotistical interpretations of the emotion."

In short, the only thing this love shares with the forms of love that are experienced by all of us in day-to-day life is the word *love* itself. This place of "perfection," as has already been described through so many people's testimonies, is imbued with goodness and compassionate understanding, and it is where people perceive being assisted to gain greater knowledge by other entities that have a higher magnitude.

5. TIME: NONLINEAR AND NOT AS IT WOULD SEEM

One of the features that people consistently report is that their perception of time changes. During surgery for a life-threatening bowel condition, an American woman, Sharon, had a recalled experience of death and explained, "I had no sense of time; I was there for what seemed like eternity." Diane, the American woman who had described being able to multi-locate and had separated and liberated from her body, as if an armor had been removed, explained, "In those seven minutes of death . . . I experienced seven weeks of time or more."

Mohammed said that during his experience, "Time had lost its meaning." Tracy suffered a cardiac arrest in 2009 and said, "[It] seemed like hours but afterwards I discovered I had only been without a heartbeat for 1 min and 14 seconds."

Laura, who had suffered a gynecological emergency, explained, "I had realized at that point that time was different and did not really matter." Joan said, "Time did not, does not, really exist." Mache described how "time wrapped in on itself. There was no past, present, or future as we see it here." She clarified, "It seemed like things were not linear." This was also echoed by Jeff, who said, "I remember things, but it's like they all happened at once. Time ceased to exist."

The issue of nonlinearity of time is an important feature. Lauren, a twenty-nine-year-old yoga teacher, suffered an unexpected cardiac arrest in 2017. She explained, "I was in the middle of teaching yoga to a few students when I just collapsed. I dropped dead, went kaput. Just like that. With the flip of a switch, my heart turned off. One student kept my heart going with CPR for 20 minutes until the ambulance arrived." She continued to explain, "My experience of the situation was quite different. It was as though I suddenly existed outside of the boundaries of time. There was no linear progression, no going from point A to point B." Heidi, who had suffered a complication during childbirth in 2004, explained, "Time as we know it does not exist there, as it does here."

Katherine underwent a life-threatening complication of surgery in 2001. She explained that "time only passed in a linear fashion because I organized the different events as happening in a certain order when I reflect on it. It's extremely hard to explain." *By this she meant that after people recover and recall their experiences, the events fall into a sense of linear order because this is what the human brain is used to doing, even though the experience itself wasn't linear.*

Emanuele, an agnostic Italian man who had a cardiac arrest after a heart attack, explained, "Time didn't exist because everything was happening at the same time. Time wasn't linear like we know it."

It is important to highlight that as these cases illustrate, things all happen at once and not in a linear manner, even if survivors describe events in a linear manner later. In this book, due to the limitations of the way that the human brain functions, we have also had to discuss the different features and themes that are experienced and recalled by people in a sequential manner.

Bobbi described this concept really well. She said, "Now this is going to be hard for you to comprehend. So I'm really going to try hard [to explain it]. Everything is happening at once or overlapping at best. Imagine having a thousand things, and people talking at the exact same moment, and perfectly understanding every detail of all that's going on. I could understand all six [entities] speaking at the exact same time with perfect clarity, as well as knowing the purest depths of their hearts, and a multitude of other things happening and information that I was privileged to receive, all at the exact same instance. There are no doubts about what is said because of the dynamics of the form of communication."

6. A GLANCE AT MY PAST AND PRIOR PAST

Some individuals realize they have had lived past lives during their recalled experience of death. Jeff, who had suffered a life-threatening asthma attack at the age of seven, explained, "It's hard for me to describe what happened next . . . but I understood it to be a recounting of this and past lives."

Telesa, who had suffered a complication of childbirth in 2017, also described something similar and said, "My understanding is that we live numerous lives, each with a different 'purpose.'" She continued, "This life is all such an infinitesimal portion of who we are. Who we are today is a result of all of our previous lives combined." Another person explained, "I knew this wasn't my first time here [referring to the place that feels like home]." He then said, "I learned that we come here . . . many, many times."

Diane shared that "I even felt as if I'd done this before and was remembering that I was going home." Anni told us, "I experienced three past lives," before explaining, "I was shown various clips from past lives. I had never before related to [this concept of successive lives]. I had grown up in a family of nonbelievers . . . "

In 1963, Teresa was three years old. She explained that one day she had sneaked into a "diaper bag and found [a] container of hidden aspirin." She said, "I ate them all. Needless to say, later while watching cartoons, my eyes rolled back, and I went into convulsions. My mother screamed for help, and I was transported to the hospital... Too much time had passed, and the aspirin was already in my bloodstream. By the time we made it to the hospital, I was limp and lifeless in my mother's arms."

She continued to explain, "The medical personnel rushed me into a room where they started trying to revive me. As they worked on me, I left my body and was watching from above. I could see and hear everything but could not feel what they were doing to the body below. I can remember vividly watching with curiosity and interest. While watching the doctors and nurses working on the body below, I also remember being able to watch what was happening at the same time in the room they had taken my mom. I could feel my mother's hysteria and distraught, as if her emotions were my own. I remember being overwhelmed with feelings of intense love for my mother." She then added, *"I had memories of a past life that I had shared before this one that involved both my mother and my sister."* She then explained, *"I became aware of a number of prior lives, from earlier times."*

Ingrid explained, "Somehow I comprehended that the current life that I was living was just a continuation of a very long existence."

Doug, a British man who was gravely ill in India, explained that during his experience "other lives lived started to appear... the entire life from beginning to end, in an instant, with awareness of every moment of that life as if it was just lived from birth to death."

Celso got into trouble swimming in Aruba. During his recalled experience of death, he, too, described an understanding of past lives. He said, "I started to recall significant moments of my past life. It was like traveling to the past to be the main character of the story and an observer at the same time."

Interestingly, we found survivors don't refer to these experiences based on traditional notions of reincarnation derived from Eastern religions or metempsychosis from ancient philosophy. In short, they don't mention it even in the way that past lives are discussed in popular contemporary culture. They don't describe being famous people or animals or refer to any religious or cultural traditions about reincarnation. *Instead, there is an understanding that one lifetime may not be sufficient to achieve the specific levels of knowledge and understanding that are needed. What is described is instead a recognition that an ascending series of human lives exist, which are meant to help people fulfill this purpose.*

7. COMMUNICATION: THOUGHT IS EVERYTHING

One of the things that people consistently describe in relation to their recalled experience of death is that all communication with other beings occurs directly through thought. Some people may use the word *telepathy* to describe this. There is no verbalization in the way that we are ordinarily accustomed to, through movement of the mouth, lips, and so on.

Wayne explained, "I sensed a presence behind me and then had a communication. This was beyond telepathy. This was not hearing words in my mind and translating them into thoughts, this was knowing as the other presence knew, an instant sharing of knowledge." Laura, who had suffered a gynecological emergency, continued to explain, "I knew I had died to the world, but [from her own inner perspective] I had not lost consciousness for even a second. I was still me and still alive." She then described meeting a luminous entity and was "enveloped in such a great light and love that it defies description." She then explained, "He spoke to me without words, without a voice, and yet I had clearly heard and understood every unspoken word! He said to me that I was in a different 'place.'" Bolette described how "I was surrounded by this unearthly loving, very beautifully radiating light. In that overwhelming radiating, loving light, I met a glowingly beautiful, very loving being." She continued, "We communicated with the use of our thoughts and mind."

An American physician, Gillian, described suffering a cardiac arrest in 2013. She said, "Time was meaningless. I was with a group of beings that I felt I had known for a very long time . . . I have a vague recollection of having my Earthly experiences 'downloaded' and having a great reunion with these beings." She then added, "Communication was nonverbal and instantaneous. It involved relaying entire occurrences, concepts, and events with associated emotions, not just words and sentences. Eventually a consensus was reached that I should return to the life I had left, as it was unfinished."

Many other people have conveyed this in different ways. It is a very constant feature of the recalled experience of death. One person said, "[The being] was communicating with me simply by thought." Another said, "No words were exchanged. Thoughts moved instantaneously, with perfect clarity, from one part of the eternal mind to another; without the ability to withhold or judge anything." One other said, "Everything was communicated to me without any words whatsoever," while another individual explained this as "the beings communicated with me and one another telepathically. We spoke without words, directly, from mind to mind."

8. A SOURCE, AN ORIGIN

Many individuals reported a sense that a source, or origin, exists. This is felt to have a magnitude that is way beyond even the magnitude of the highest luminous beings of light. It feels so high that it is unknowable and imperceptible, unlike those guiding beings of light that can at least be perceived. Romy, a young woman whose heart stopped after a traffic accident, described how "I had never felt it so strongly. It was everything. Everything I have ever needed . . . [it had an] immense healing and nourishing quality to it. It was pure, immense, powerful, unconditional love. I knew I could trust this light . . . " Wayne, an older American, described how "I was made to know there were an infinite number of realms of existence and all were part of the One, the Source." Garca explained, "I wished to go to the source that emanated so much love . . . I wanted to reach it." Telesa said, "It was as if I were being shown that the central sun was where all energy comes from and returns to, no matter where one calls home from lifetime to lifetime."

9. FEELS MORE REAL: ORDINARY LIFE IS LIKE A DREAM

As already seen, many of our interview subjects shared a sense that this experience was hyperrealistic. Lavette described how "I was amazed at how much more real and vibrant the colors and light around me were." Emanuele said, "Our daily life seems like a dream in comparison to my experience." Another subject recounted, "I know what I experienced was not a dream." Telesa had a similar understanding. She explained, "Everything was hyperrealistic, perhaps more real than I have ever known reality to be." Jeanne, an older French woman, described it as "colors, sensations, feelings that are so far above anything we have here it is as if a veil has been lifted." Michael's recall was that "It wasn't dream-like because everything was in clear detail." And Anna reported that "the one word I'd address to the entire experience would be *reality* . . . that place, that event were the most real thing that's ever happened to me."

Overall, Karen explained it best when she said, "I viewed it as a true experience, because I experienced it. The same way I am experiencing my sitting here at my table on my balcony and typing these words, hearing the TV on a program to which I'm not paying attention. *How do I know what I am experiencing right now is real? I just do. I know the difference between a dream I have had the night before and the reality of waking up to my day-to-day life. How? I just do.*"

10. LEARNING THE IMPORTANCE OF MAKING AMENDS FOR INFRINGEMENTS

Survivors with a recalled experience of death often realize that they need to take accountability for their actions in life and try to make amends for the wrongs they have done to others, and that this is itself meaningful. This is exemplified by Keri, a young woman who reported a feeling that *"he [the entity] saw if I was sorry for the bad things I had done, how I felt in my heart at the time I had done things, how I felt afterwards, and whether or not I apologized or made the bad things right."*

11. ACQUIRING KNOWLEDGE OF THE FUTURE

To help them better understand the higher purpose to life, many respondents also saw portions of their future. One young man shared, "I was shown parts of my future life, like going up to a screen and suddenly being in the moment experiencing it. It was as though I was there at that moment, feeling how I would feel at that time. I was shown parts of my future in this life if I chose to go on." We heard many different versions of this sentiment, including the idea that people recognized they would return to "this place upon the conclusion of my life. I would go through another review process, and I would stay there until I wanted to come back . . . " Some individuals had a sense that they had seen images of the future but that they were erased when they were resuscitated back to life.

12. SUICIDE: NOT AN END AND NOT AN OPTION

Most people enter the grey zone after encountering a life-threatening situation or illness, but some who go there have survived a suicide attempt. This is a complex issue. For one thing, those people who return after a suicide attempt are clearly different from those who do not. As we have seen already through many testimonies, one of the hardest issues that people face in the grey zone of death is reexperiencing the same pain and distress that their actions may have caused others, as well as the downstream consequences of their actions on the lives of others. Therefore, the person who successfully commits suicide will likely experience a very different level of anguish—considering the effect of their actions on others—compared with those who do not.

One of our testimonies was from an accountant with a wife and three small children, whose marriage was falling apart. He explained that he had been consumed with fear and self-pity and had become an alcoholic, which had, of course, further wrecked his marriage. One night he decided to end his misery by taking his life using numerous sleeping pills with alcohol.

He described how "I felt myself moving through this tunnel at a very fast rate. I saw a light at the end of the tunnel and was wondering if this is where I was going. I didn't know if I was dead or alive at that point, but I do recall looking back at myself passed out on the kitchen floor, and I lay there completely oblivious to this other part of me which seemed to be heading toward something. 'Is this what death is?' I wondered. 'No!' came an answer from somewhere."

He continued to explain, "I was shocked to see a being of incredible beauty, radiating great love, great compassion and warmth . . . I was hesitant to say anything, and then I realized that my thoughts were being read by this incredible being of light. 'No!' he repeated again. 'This is not what death is like. Come, I will show you.' I remembered floating with him over to a pit of some sort that contained a very depressing scene of a landscape devoid of beauty, devoid of life, where people shuffled around with their heads down and their shoulders bent forward in a depressed, resigned manner." *This contrasts so much with the different types of luminous beings, including happy family members that we have seen other people describe encountering during the experience so far.* He continued to explain, "They kept their heads down and looked at their feet and wandered around aimlessly, bumping into each other occasionally but they kept on going. It was a horrifying thought that I was going to be cast down with these confused lost souls, but the voice seemed to understand my terror and relieved it with the following words: 'This is a Hell of your own creation. You would have to go back to Earth eventually and experience a new life all over again faced with the same difficulties that you faced in this lifetime. *You will stay with these lost and confused souls until then. Suicide is not an escape.*'"

He said, "[Then] I was shown a panoramic view of my life. The last five years which had become so burdened with alcohol abuse were the most painful things, the most painful memories."

Afterward, he explained that during the experience he was shown the consequences of what would happen if he were to continue his drinking habit and in particular if he were to die. He too was shown the future. He saw that his wife would fail their children, and that they would end up in foster care, and eventually ruin their own lives with alcohol. He continued, "It was like a slap in the face. A huge reality check. I saw that if I shaped up my act and began behaving like a responsible father and role model, all three children would grow up to be happy and productive. That doesn't mean completely free of the struggles of everyday life, but they would have a chance at making their own way, independent of any substance abuse . . . "

In death, people recognize that they cannot truly relieve themselves of whatever had been causing them to suffer through an act of suicide, because their consciousness and selfhood is not annihilated. They learn that whatever life challenge they may have been facing, and as hard as those may have been, there is no real escape from them. In other words, even if they do manage to commit suicide, their pain and the challenge they needed to overcome won't end.

People who die by suicide, unlike the other cases we have reviewed, experience a miserable state that they feel to be a hell of their own making. They realize that they have to live with the painful effects of their actions on others, which will add to the misery. They also discover that eventually they will be sent to live another life with the same challenges that they were trying to escape from. In short, it seems like instead of escaping from hardship through suicide, the message that people come back with is of the need to try to overcome and dominate the challenge internally and learn, rather than being dominated by it. That is because ultimately, as Dr. Mary Neal had described, even in the hardest moments in life, there was an opportunity to gain insights and understanding, which she referred to as the "beauty" in those situations. Overall, gaining knowledge through each challenge seemed to be what was sought from those challenges in life. There was also a recognition in her experience and many other people's testimonies that a reason—cause and effect—underlies all events in life. That also explains why there is no real escape through suicide, because it is not possible to break the chain of cause and effect that links everything together.

HAVING GAINED SO MUCH INSIGHT AND KNOWLEDGE, THE FINAL stage of the experience is a return to the body, and to life, as people recognize they have a second chance to pursue the higher purpose they have come to appreciate existed in life. This experience leaves people with long-lasting positive transformative effects.

This phase of the recalled experience of death can be divided into at least six distinct components or subthemes:

1. *A desire not to return.*
2. *A point of no return.*
3. *Decision to return.*
4. *Perception of return to the body.*
5. *Back in the body.*
6. *Fragmented memories and long-term aftermath.*

1. A DESIRE NOT TO RETURN

Overwhelmingly, survivors talk about not wanting to return, about wanting to stay where they are, and of being unwilling to leave a place that feels more real than real life to them. Some are presented with a choice, but others describe a feeling that they must return to their responsibilities in life. Lara, a middle-aged woman, described that she knew she didn't want to go back. "Nothing would ever be as perfect . . . as this." Another woman described how "a being came to me and asked me to return to the body. I was reluctant to do so." Finally, a male subject described feeling insulted at the idea that he might want to go back. "I was horrified at the thought and felt myself loud within me respond 'No!' There was a pause, and I felt a little confused . . . Again, the same question repeated with me. 'Do you wish to go back?' Again, I said, 'No.'"

2. POINT OF NO RETURN

Returnees sometimes report one final moment of choice. "I came to another portal and my aunt [who had died four months ago] told me, 'If you pass through here, you will not be able to return.'" Another reported being aware that "you cannot cross this barrier. It is time for you to go back."

3. DECISION TO RETURN

In the end, some individuals chose to return for the sake of their family—even though many initially felt a sense of disconnection from their loved ones at the beginning of the experience. Sarah, a young American mother, recounted how her aunt appeared to her during her recalled experience of death. Her aunt was giving her a choice, and momentarily Sarah didn't feel there was anything to return to. Eventually, however, "[I] heard my daughter's voice . . . She asked me to come back. I made the decision to return." Others worry about leaving their spouses or parents.

A variation of this is a sense of being asked to return. "Somehow it was conveyed to me that it was not my time yet, and I had to return."

Some described a sense of needing to return to fulfill tasks or develop as a human being. Alana articulated that "there was so much that I had to do. I had to improve as a human being." Another person shared that "I saw that I had to return . . . in order to finish many things." It is not an easy choice; many shared that they felt a sense of obligation, either to themselves or others. Emma said that "I had to return to my body . . . to face my fears."

4. PERCEPTION OF RETURN TO THE BODY

The final element of a recalled experience of death is the return to the body. Unlike the pain-free nature of the experience itself (from a physical perspective), returning to the body is unpleasant. The initial feelings of the return are frequently described as difficult and intense. One interviewee described "shooting backwards and it was cold and dark, and I was grieving the light before I even hit the body." Another shared how "I felt myself being pulled back and away from the light. I did not want to go and recall feeling sad."

5. BACK IN THE BODY

Finally, survivors with a recalled experience of death describe the sensation of being "slammed back into [the] body . . . [I] felt intensely everything they were doing to bring me back. I had four IVs inserted in various places through my body as the nurses injected me three separate times." Uniformly, people shared that the experience of returning to the body "wasn't pleasant." Larry, an older American, shared, "I did come back and that was really cruel because it felt as if I was strapped into a straitjacket after I had been able to move freely."

6. FRAGMENTED MEMORIES AND THE LONG-TERM AFTERMATH

Survivors with a recalled experience of death are deeply altered by their experiences in the grey zone. They describe *long-term positive transformational experiences,* including a desire to "be of service to others." As they move through life, they describe a newfound awareness of the need to be compassionate to others and to make better choices in their own lives. One person said, "When I saw how impure I was, I realized that we must not condemn other people, no matter who they are." People also start to be "more attentive to the needs of people, animals, and plants. I wanted to make good, forgive people." Another described how "it's easier for me to put myself in other people's shoes." Many described a newfound sense of empathy and altruism. Overwhelmingly, individuals express a desire to be of service, be better, and to do good. One man shared how he now realized that "chasing both fame and fortune with a win-at-all-costs mentality [is meaningless] . . . Now being kind and affectionate to others makes my life full of joy." So many find a different and higher overall purpose to life.

Survivors share a sense that there are big lessons to be learned from hardship. One wrote, "I feel that our suffering is the greatest of all blessings. We learn the most from that which nearly kills us. Each person's hardships are directly related to the lesson that person must learn. It is our choice whether we accept the challenge or not." However, while survivors emerge with a clear sense of some kind of "God" existing, they do not link what they have experienced to any particular religion or institution. "Today I believe only in God, not in churches as institutions anymore." As already seen, they recognize that irrespective of what people may want to call it and what name they attach to it, there is a source, which is infinitely higher that anything they can perceive.

Overall, though, as powerful as the experience is, what is recalled is a pale reflection of what had been experienced. In the grey zone, people recall everything that had been stored in their consciousness (even if they couldn't in life); *this suggests the full experience is retained, just unable to be retrieved.*

As mentioned, this is likely because memories are lost after the liberated state of vast consciousness is forced back from what is perceived to have been a multidimensional reality, into the limitations of the three-dimensional brain and body. In addition, there are at least two other reasons. After cardiac arrest, the brain is impacted by severe inflammation. This, along with the effect of sedative drugs, affects memory circuits in the brain. In AWARE-II, we found that almost 40 percent of people had a perception of awareness, but most couldn't recall any specific features. This is also why so many survivors don't have any recall of their experience with death, whereas others only have fragmented memories. *The absence of recall doesn't mean there was an absence of experience. These should be considered as two separate issues.*

Many people's reports highlighted their memories quickly fading. "My understanding, now back in the body, is hugely limited once again." Another shared, "I knew so much instantly that I don't remember anymore at this time. I can only speculate that I can't know it in the physical state I'm in now."

For decades, testimonies by people like Dr. Montgomery were systematically dismissed as either imaginary, hallucinatory, or even fabrications by a handful of doctors or scientists without actual scientific evidence. As we shall see in chapters 15 to 17, this tendency for dismissal, without any scientific evidence, simply reflects either the inherent tendency by people to reject new phenomena that counter their own personal beliefs or their attempts at trying to fit them into previously understood models, even if they don't fit at all.

The experience that survivors describe is, by nature, hard to quantify or distill into purely conventional scientific terms. And it hints at something that most people, even scientists, have taken for granted up until now, a theory that no scientist has ever been able to prove or disprove, though many, including numerous Nobel Prize–winning scientists, have certainly tried: the origin of consciousness, or what the ancients called the seat of the soul. This unresolved story of consciousness—*of the essence of who we are*—is today considered by many to be the biggest mystery in science.

Section Three

EXPLORING CONSCIOUSNESS

Chapter 13

A NOBEL PRIZE-WINNING IDEA

On December 10, 1963, the eighty-one-year-old Swedish king, Gustaf VI Adolf, and his wife, bejeweled and in her crown and royal finery, entered the glittering Stockholm Concert Hall to a rapt audience. Several royals and dignitaries followed them while escorting their assigned honorees to their places. Every seat on the auditorium's main floor and two tiers of surrounding balconies were filled with dignitaries, scholars, and world leaders who had arrived from all over the world for this annual event. The men in the crowd, wearing tuxedos with coattails, and the women, dressed in their finest gowns and festooned in jeweled bracelets and elbow-length gloves, whispered to each other in hushed yet eager voices. The anticipation in the room was palpable. The room shimmered and sparkled as people moved about and settled. As the lights dimmed on the crowd and people looked toward the stage—no music played. No curtain lifted. Yet everyone was spellbound as Professor R. Granit stepped onstage. Taking the podium, he greeted all in attendance, "Your Majesties, Your Royal Highnesses, Ladies, and Gentlemen."

You could hear a pin drop.

Wasting no time, Granit announced the awarding of the Nobel Prize—the most coveted prize for human achievement—to a trio of scientists and researchers: Alan Lloyd Hodgkin, Andrew Huxley, and John Carew Eccles. Their work looked at how brain cells communicate and how messages are transmitted from the brain to the rest of the body through a network of brain cell connections. They studied the ionic mechanisms in the peripheral and central portions of the nerve cell membrane. While Hodgkin and Huxley's contributions are significant, I have found Sir John's research particularly intriguing, especially how it relates to understanding our minds, understanding consciousness, and understanding ourselves. Like Nenad Sestan and many other scientists, Sir John, too, was fascinated by what makes us all uniquely human and different from, say, a chimpanzee.

Eccles came upon his great discovery when he tested whether nerve cells communicated via electric processes. When he stimulated nerves, he detected what are called excitatory (meaning *activating*) postsynaptic potentials (EPSPs) in individual nerve cells. This allowed messages to get through from one nerve connection to the next along the brain or spinal cord. To his surprise, when Sir John induced an inhibitory state in the nerve cell pool—meaning a way to prevent the message from getting through—no messages were transferred. They were just mirrored: he could not prove the hypothesis. This led him to change his theory, and with a new set of experiments, he ultimately showed that synaptic transmission—meaning the process through which one brain nerve cell transmits its message to another through connections between cells (known as *synapses*)—*must* involve a chemical intermediary process.

Sir John's experiments showed how a nerve cell that has been "switched on" by receiving a signal passes that signal or message as an electrical charge on to the next cell in line. The process occurs when minuscule openings in the cell membranes expand, whereby ions—electronically charged elements like sodium—can flood inside. Once inside the cell's walls, their arrival reverses the cell's electrical charge. This process keeps happening, over and over, through a chain reaction of cells, allowing for the transmission of the original electrical impulse through the body. He also varied the process to see how it is possible to block messages and "turn" transmission through the nerve cells on and off.

Eccles won the Nobel Prize because he showed how chemical intermediaries are vital in passing electrical signals from one nerve cell to the next. In short, he discovered how brain cell connections work and how signals or messages are transmitted and either sent through or stopped. His revolutionary discovery about how messages travel from the brain has changed science and medicine forever. Eccles was one of the most prominent scientists to study the nervous system on the molecular level, and he is widely viewed as one of the greatest neuroscientists of the twentieth century.

Eccles never lost sight of the big picture. *What did it all mean? What did all these transmissions through neural pathways add up to?* As someone who knew the importance of testing one's hypotheses and beliefs, even if they might be proved wrong in the process, Sir John was persistent. He never gave up trying to answer questions that have stumped philosophers, religious leaders, and scientists alike: *Where does the human mind and psyche reside? What is the nature of consciousness? Where do our thoughts, mental activity, and our sense of self—our selfhood—come from?*

For the next two decades, Sir John continued his work on the nervous system and how it communicated on the ionic and molecular level throughout the brain and spinal cord, collectively called the central nervous system. He studied the major areas of the brain: the hippocampus, thalamus, and cerebellum. As the world's foremost expert and a Nobel Prize winner on brain cell connections, he was particularly interested in exploring what many of the greatest thinkers and philosophers had also tried to explore: the "mind–body" problem and what is also referred to as the "problem of consciousness." Sir John understood more than anyone else at that time about the material and chemical transactions of the brain, yet the more he understood about those processes, the more *he was baffled as they just did not add up.* How can those electrical and chemical transactions possibly result in who we are? Why are we so unique? And where do we get our thoughts, intentions, feelings, and personality from? These are the things that make us all human. He asked: Is our mental world and our sense of self just simply the sum of trillions of nerve impulses and transmissions? Is that all there is to us?

Eccles's early work was groundbreaking. It earned him the honor of a knighthood from Queen Elizabeth II and a Nobel Prize. Still, Sir John was unsatisfied with his research. For example, although his discovery pointed out that synaptic transmission included a complex quantum-mechanical chemical element (not simply electrical), he had a hunch that there was something more behind what he and his colleagues had discovered.

In 1966, he moved from Australia to the United States to continue his research on the brain. Sir John was particularly interested in the work of another scientist, the American Roger Sperry, who was performing groundbreaking studies on animals and patients whose brain hemispheres (meaning the right and left halves of their brains) were disconnected (this work later led Sperry to a Nobel Prize in 1981). In addition, Sir John continued to be intrigued by the origination of different thought processes, speech, and consciousness in the brain. With Sperry's experiments, he began a quest to find what has eluded scientists, philosophers, and doctors for millennia: the source of consciousness. "Brain research is the ultimate problem confronting man," Sir John argued. "To the extent that we have a better understanding of the brain, we will have a richer appreciation of ourselves, of our fellow men and society, and in fact, of the whole world and its problems." However, he added, "I can explain my body and my brain, but there's something more. I can't explain my existence—what makes me a unique human being." This is the question that has baffled philosophers and scientists alike for millennia.

Some sixty years later, it is a question that still baffles us all.

DR. MEGAN CRAIG IS A SOFT-SPOKEN ARTIST AND PHILOSOPHY PROfessor at the State University of New York in Stony Brook and author of *Levinas and James: Toward a Pragmatic Phenomenology*. She, like many others, has had to confront the same questions that had baffled Sir John. In Megan's case, though, the need was thrust upon her, without much choice, after an unfortunate and unforeseen accident.

In January 2018, Dr. Craig chaperoned her daughter's school ice-skating trip. While she focused on her group, a sturdy little boy of no more than nine years old lost control of his skates and slammed into her from behind. "He was just the right-size projectile to undercut my skates and send me flying backwards on ice, where I landed on my head," she explained. Immediately after standing up from her head injury, Megan noticed that she was not her usual self. "I was disoriented and could not remember my address, or the date." She was taken to the hospital and diagnosed with concussion. The doctor told her she would have to undergo "brain rest," meaning no reading, no writing, and no computer screens, and, most importantly, no strenuous thinking for three months. For Megan, "being a philosopher on brain rest" was "like being a point guard [basketball player] on hand rest." Her greatest asset, her brain, was suddenly not working reliably. The prescribed treatment, brain rest, was elusive, because like everyone else, Megan was incapable of "arresting thought." She later wrote of her ordeal in the *New York Times* and explained that, unlike with other injuries, "there was no protective cone to keep the wounded place from being itched, no cast to keep the brain still."

For Megan, the philosophy professor, "brain rest inspired more thinking about the very nature of thinking." Her training had offered her a deep understanding of the philosophy of mind, the branch of philosophy that deals with what Nobel Prize winners like Sir John Eccles, but also other profound thinkers throughout history (Plato, Aristotle, Avicenna, Descartes, and more recently William James and Carl Jung, among others), had contemplated. From one moment to the next, Dr. Craig had unwittingly become the living embodiment of the problem that had baffled them all. She was going through an experience of mental "dislocation," which led her to question things she had taken for granted as a philosopher. In particular, she said, "the degree to which I am or am not my brain" and what constituted a whole person. "Not having access to memory in the way that I had before was extremely disorienting and really made me question to what degree I have personal identity if I don't have memory."

"It was also very isolating, because I would be trying to communicate, or be with even the people most close to me and who I love, and not be able to find their name in the inventory that I knew was there somewhere, but I just couldn't find it." In the days after her concussion, against her doctor's orders, she tried to muster all her energy to read her daughter several pages of a storybook, only to fail to reach the end. This made her feel totally embarrassed and demoralized.

Her condition got worse over time. As she lived in this frightening state of mental "dislocation" for days, weeks, and months, she found herself becoming much more empathetic toward people with any kind of experience of dementia or brain trauma. She recalled her interactions with an elderly woman with Alzheimer's disease, whom she had cared for when she was in high school. "Every day I would show up at her doorstep, and every day she would ask, 'Who are you?' and we would go through this whole ritual of meeting each other all over again." But now Megan started questioning what was really going on in that woman's mind. Had she, like Megan, been experiencing herself as someone who was independent of, but trapped and tethered to, a malfunctioning brain apparatus, or was she—her selfhood, mind, and consciousness—the malfunctioning brain itself?

Megan, a philosopher and a pragmatist, started to question to what extent we humans are our brains or are separate from our brains. After several months, she eventually recovered from her terrifying ordeal. She found herself capable of accessing her memories and recalled how her own grandfather with dementia, late in his life, had described "his own failing mind as a library where all of the books were shelved too high to reach." Megan had experienced this at firsthand—*as a separate observer*—with her brain and its disorder. Her experience left her more acutely aware of the predicament and suffering that people with dementia or other brain disorders go through.

Perhaps, as Megan had experienced, the old woman with dementia, her grandfather, and other people with brain disorders retain their selfhood and mind, yet can't access their memories, or other functions, because their brain apparatus is faulty. "This is what dementia must feel like," she thought. *Maybe consciousness, selfhood, and the mind do not disappear in people with brain injuries or diseases like dementia.* Perhaps instead, just as Megan had experienced firsthand, people maintain their independent and separate sense of self and cohesive mind, yet are trapped by a brain that isn't working properly. They find that they can't access their memories, including even the simple task of remembering the names of loved ones, or other brain functions, all while their mind and selfhood is able to observe the limitations of their own brain. This would be like a person who is unable to walk but observes their own nonworking body. Yet with patients with dementia and other brain-based disorders, people mistakenly think that the patients' mind and selfhood is absent because of how their brain malfunction makes them appear and behave. Their behavior suggests that their "self" is gone, and the outside world dismisses them. But that may not be true.

Megan explained, "It's very easy from the outside to make assumptions that somebody is not with it or not bright or not articulate, when in fact they may just be unreachable in some way. That injury really made me think about all kinds of instances of injury, and especially different kinds of mental illness that have those same isolating effects and that we tend as a society to be so quick to judge about."

For Megan, concussion had "a way of changing one's sense of the balance between mind and body." In her *New York Times* article, she wrote, "It's one thing to be hit in the arm or the gut. It is an entirely different thing to be hit in the brain . . . The injury to my brain highlighted the degree to which my identity and my powers of identification have a specific seat in my brain. The concussed condition was an intimation of how terrifying dementia and other brain disorders must feel; the loss of a thread that has so far tied together one's life and tethered it to the lives of those one loves. In my case, the loss was temporary, but it was the first time that I have ever felt so distinctly the efforts of thinking, the brain as a muscle and the depression that accompanies the feeling of having lost oneself. I thought about memory and the loneliness of being unable to recall names and places that are the road marks of a communal life. Though my condition was transient, it highlighted the degree to which the brain serves as an anchoring center of control. For months, uncertainty accompanied my every move." Overall, the experience was enlightening, as "for the first time in my life, I really felt like I was kind of a witness to my brain, *but not located in my brain.*"

MEGAN'S FIRSTHAND EXPERIENCE AND THE QUESTIONS THAT AROSE from it have baffled philosophers and scientists alike for centuries. Is our sense of consciousness and our awareness, as well as our mind and selfhood, a biproduct of bodily processes, or is it an entity that functions independent of the body, while intimately interacting with it? Today, many philosophers and scientists continue to debate the "mind–body problem" and the "problem of consciousness." Ultimately, we are all thinking conscious beings. All our actions come from our thoughts and intentions. Yet, as noted by Eccles, the substance of thought—*including consciousness and awareness itself*—seems fundamentally to be neither the same as the electrical nor the chemical process that he had discovered transmit messages throughout the brain. It is very different.

These questions are what some modern neuroscientists, including Dr. Nenad Sestan, have also tried to grapple with. Throughout his efforts to study the brain, Sestan, too, continued to bring us back to his original intent in studying those postmortem pig brains. "I wanted to understand what makes us human. No animal can write poetry," Sestan said. "If I find out what makes us human, I would die a happy scientist." He added, "There are two great mysteries of science . . . [The first is:] Who we are and what we are as a species. [The second is:] How you get thought in the brain."

I would add that there is a third great mystery in science, and in life, and that it is the natural successor to our question of self and thought: What happens when we die, and does our "self" persist in any recognizable form after our death? To answer the question of why there is a hyperconsciousness in death, we must first have clarity about how and where human *consciousness* comes from, and what mechanisms create it in the first place. The AWARE-I and -II studies have hinted at the astonishing possibility that consciousness and the "self" are not annihilated when we cross into the grey zone of death. Instead, the self and the consciousness is liberated through the mechanism of disinhibition. People gain access to what had been a vastly larger entity of their own consciousness, which in this seemingly liberated state can take in information in the form of knowing, even though normally it would not be possible for the dying or dead human to see, hear, or witness events. Yet people consistently report it. This suggests human consciousness is like a field of energy (like electromagnetism) that is tethered to the body and brain and interacts with them, yet is not produced by them. This is an astonishing proposition, but less so when you consider the sheer power of a human brain: an unfathomably complex organ that we are still exploring and charting, like an unknown ocean or solar system, together with the mystery that is itself like a universe, human consciousness.

Sestan made the point that we humans are not unique in having complex, large brains. Yet we are alone in having culture, science, language, and arts. Human consciousness contains unique features unheard of in any other living being, even the smartest ones in the animal kingdom. As he put it, "There is no doubt that we can do things that chimpanzees cannot." Humans, unlike the great apes or other primates, learn complex languages and compose poems rich with metaphor and emotion. Our power of creation is unparalleled. We launch ourselves to the moon and voyage to the bottom of the oceans. We compose astounding symphonies and cure diseases. And we pass on our science, our art and culture, and our knowledge to the generations that come after us. We, unlike any other creature on this planet, have a drive for intellectual complexity and artistic endeavor. Importantly, we also strive toward morality and ethics. And that sets us apart from the animal kingdom.

For Sestan, "There are two important things to understand about what makes us human. One is that we have a bigger brain [than most other animals]." But the fact that "our brain is three times bigger than chimpanzees' is not exceptional." In fact, he explained most scientists wouldn't even be able to tell the difference between a chimpanzee and human brain. To top it off, "there are other species that have bigger brains." So, it isn't the size of the brain that distinguishes us from other animals, especially since there are some humans with very small brains who have normal intelligence and function.

There are plenty of humans who survive and thrive with severe brain abnormalities. People born with severe schizencephaly have only half of the brain. There are also many examples of children and infants who needed to have half, a complete hemisphere, of their brain removed. Yet many are able to live normal lives, as the remaining hemisphere compensates for the lack of its other half. Individuals with hydrocephaly have a brain that is reduced to a sliver, yet they still function normally. In 2007, a prestigious medical journal, *The Lancet*, reported on the baffling case of a forty-four-year-old French civil servant who had gone to a hospital complaining of leg weakness. When doctors performed a brain scan, they found he was missing 90 percent of his brain. This was a man who was living a completely normal life, with a family, stable job, and a massive brain abnormality that was found by total chance.

Yet in all these people, even the Frenchman who was missing 90 percent of his brain, the sense of selfhood and consciousness continues to exist. It isn't diminished in proportion to the relative lack of brain mass. Normally when we experience a conscious feeling or any thought, multiple areas of the brain become active at the same time and work in unison together. However, if the mind and consciousness in someone without 90 percent of their brain—including much of those regions ordinarily involved with conscious thought processes—can still function normally, then we can't assume that our conscious experience is produced by those specific parts. Why not? Well, because most of those brain parts were missing. Instead, perhaps the mind and consciousness may be some other property, and that is why, even with a sliver of a brain left (if the brain has had time to adapt over time, rather than undergoing a sudden massive injury that might kill off large chunks of it quickly), it is possible to mediate normal conscious activities. The key here is that mediating consciousness is not the same as producing consciousness.

Either way, certainly size is not the issue. Whales have bigger brains and are clearly intelligent. But they are not humans. Likewise, dolphins are very intelligent, but they have not created the highly complex societies and cultures of the human race. They cannot conquer diseases or find ways to live on land, let alone travel to space.

If it is not the size, then perhaps it's the connection between our cells. This is why Dr. Sestan studied postmortem pig brains. In his view, "the reason you have human cognition is not necessarily just the size—it is that our neurons are connected slightly differently. Maybe we have more connections, or we have connections that other species don't have."

While Sestan is clearly an accomplished and highly respected neuroscientist, a theory based on the belief that humans may have different brain cell connections still does not address the burning question: How can a single thought or the sense of being conscious arise from chemical and electrical processes? What about the fact that some simple organisms and some simpler animals do not have any brain cells? Can we assume they are not conscious? Also, how can a person with just a sliver of a brain and far fewer connections still maintain the same uniquely human sense of consciousness with all its incredible abilities, rather than descend into a lower level of conscious abilities, as we might see with some animals that cannot do what humans can?

If we peer down a microscope at a single brain cell, the problem that baffled Eccles becomes clearer. These cells, like other cells in the body, produce protein-based products. If somebody were to claim that a brain cell is now thinking, *I am feeling tired or hungry*, most of us would agree that this sounds impossible. After all, this is a single cell. Now, if we connect hundreds, thousands, or even millions of these cells together to communicate using electrical waves through the long tracts of wiring (the axons whose mechanism of action Sir John discovered and that Dr. Sestan studies), the fundamental question remains: Why would those brain cells and their connections, no matter how well they may be fused together, suddenly be able to think human thoughts? For instance, why would cells and their connections feel guilty, or worried, or happy? Why would they generate worrisome thoughts about a promotion, or moving to a new house, or making the last train? Sir John Eccles's point was that this remains totally inexplicable if we assume the brain produces our consciousness and thoughts.

In response, some philosophers and scientists have argued that a person's selfhood and their consciousness must somehow emerge from the activity of the brain, like heat coming off a fire, almost like magic. It just happens. This is why they call it an emergent property. But unlike heat, the world of consciousness and thought in humans is enormously complex. It is like a universe. Our selfhood is not just a "property" that emerges, like heat from a fire. Imagine minds and consciousnesses as vast and powerful as those of Beethoven or Mozart, capable of creating extraordinary music, or Shakespeare, capable of creating great works of literature, Rumi capable of creating poetry almost 800 years ago that continues to resonate with and touch the human heart today, or Einstein, capable of unearthing the hidden secrets of the universe. Surely these complex entities, human minds with such great power and creative genius, with all their unique features and facets, don't just somehow emerge happenstance.

While Sestan acknowledged that the whole theoretical field in neuroscience tries to deal with this, he also admitted that we still have neither a good picture nor a theory about how thought and the universe of human consciousness with its unique features arises from the action of nerve cells.

In an article titled "Theories of Consciousness," published by the prestigious scientific journal *Nature Reviews* in July 2022, Dr. Anil Seth, the director of the Center for Consciousness Science at the University of Sussex in England, highlighted a "steadily increasing focus on the development of theories of consciousness." To this end there are innumerable theories of consciousness, including: higher-order theory (HOT), self-organizing meta-representational theory, global workspace theory, neural Darwinism, beast machine theory, "self comes to mind" theory, multiple drafts model, and orchestrated objective reduction theory. These theories represent only a fraction of all the theories proposed so far by scientists and scholars to address the problem of consciousness. His article focused only on those that consider consciousness as an emergent property, despite the absence of supporting scientific data.

In general, in science, innumerable theories usually *means scientists simply do not know*. The more theories, the more uncertainty. By way of comparison, we do not have tens or hundreds of theories to explain what causes a heart attack. That is because we have data from scientific studies. However, to date, it has been impossible to find any scientific evidence whatsoever or even a plausible theory to explain the occurrence of the mind and consciousness—our real selfhood—anywhere in the brain, or through any brain process.

Some of these theories propose that mental states, meaning the aforementioned complex and deep minds of Einstein, Beethoven, Mozart, Shakespeare, and of course each of us, arise either from specific patterns of activity within networks of brain cell connections or from a specific pattern of synchronized and rhythmic electrical activity in networks—millions or billions—of brain cells connected together. But again, as Sir John Eccles pointed out, the complexity of connections cannot explain how the unique world of the human mind and consciousness comes to exist.

To simplify this, some say consciousness emerges as a novel property out of the complexity of brain cell activity, or when a critical level of complexity is reached in networks of brain cell connections. They argue that while connections between a handful of brain cells may not be sufficient to give rise to thoughts, feelings, and conscious awareness, these characteristics that define us all as human arise—again, almost as if by magic—when enough cells connect together.

Stuart Hameroff, an anesthesiologist at the University of Arizona, and Roger Penrose, a mathematician from the University of Cambridge and another Nobel Prize winner, have highlighted the many limitations of these theories to date. They argue that these theories concerning brain cell connections cannot explain the observed features of consciousness.

Instead, they have put forward another brain-based theory using quantum physics. Their theory is based upon the principle that there are two levels of explanation in physics: the familiar, classical level used to describe large-scale objects, and the quantum level used to describe very small events at the level of existence so small it is even smaller than an atom (called *the subatomic domain*).

Hameroff and Penrose propose that consciousness may arise from tiny, tubelike structures made of proteins that exist in all the cells in the body, including brain cells, and act as a skeleton that allows cells to maintain their shape. They propose that these small structures are the site of quantum, or subatomic, processes in the brain due to their structure and shape. They argue that consciousness is thus not a product of direct brain cell–to–brain cell activity, but rather the action of subatomic processes occurring in the brain.

Their argument is based on the fact that at the quantum, subatomic level, superimposed states are possible. This means that two possibilities may exist for any event at the same time. However, at the classical level, either one or the other must exist. So, although there may be subatomic processes going on at any one time with the potential of different superimposed states, when we make an observation, those superimposed states must collapse into one. In support of their theory, they further argue that there are single-celled organisms, such as amoeba (there are other larger organisms, too), that despite lacking brain cells or any connected assemblies of specialized brain cells seem to have consciousness and can swim, find food, learn, and multiply through their microtubules. Hence, they suggest this may be a more advanced structure leading to consciousness.

Other scientists have rightly argued against this theory by pointing out that microtubules exist in cells throughout the body and not just in the brain. Also, there are drugs that can damage the structure of microtubules but appear to have no effect on consciousness. More importantly, it has been argued that although the theory may potentially account for how the brain carries out complex mathematical problems, it still fails to answer the fundamental question of "how" thoughts, feelings, emotions, and what makes us into who we are arise. It just moves the problem from the level of brain cell networks and connections to tiny tubes inside cells. Yet the fundamental question and problem remain the same: How can any brain-based physical process lead to a single thought, or a sense of being conscious, or the feeling and experience of love? And how might this apply to our experiences in the grey zone and after?

SOME SCIENTISTS HAVE TRIED TO SOLVE THIS RIDDLE BY IDENTIFYing a correlation or relationship between measurable brain-based activity and conscious experiences (such as seeing an object, thinking, or having specific feelings). This is typically carried out using special brain scanners called fMRI (functional magnetic resonance imaging) and PET (positron emission tomography).

Brain cells have a constant need for blood and vital substances, such as oxygen and glucose. These scanners work by detecting the relative changes in the flow of blood or uptake of glucose in areas of the brain in response to conscious states. These measurements provide an indirect way to determine biological activity in the brain. These changes are converted into different colors and illustrated as a "heat map." (Much like the heat maps that weather forecasters use.) Typically, red is used to show areas with more temperature (more activity) and blue areas with less temperature (less activity). Similarly, brain areas with greater cell activity (hot) and less activity (cold) can also be depicted in response to a given emotion or feeling or different mental states at any time. This is called "mapping" the brain.

The markers, or signatures, that are identified are called the neural correlates of consciousness (NCC). Some scientists and the media, especially in the 1990s and 2000s, initially reported that the discovery of these brain signatures meant scientists had discovered the "seat of the soul," or consciousness, in the brain. Further, they said that changes in blood flow or glucose uptake meant that consciousness is clearly what they detected. However, others pointed out that just because one thing correlates with—is related to—something else, doesn't mean it has caused it. This is a basic law of studying correlations, or relationships, in any field of science. Correlation is not causation.

When any correlation is observed between two events, there are three possible explanations. Let's take the example of two actions, A and B. If there is a correlation, then either A is causing B, B is causing A, or some other process is causing both. So, when correlations are found between brain-based events and conscious experiences, these are the three possibilities. While we know that brain-based events show a relationship with thoughts, no one has been able to demonstrate whether A causes B, B causes A, or some other thing causes both. In other words, perhaps the brain-based events cause conscious experience, or conscious experience causes the brain-based changes (the detectable signatures of consciousness), or something else causes both.

Donald Hoffman, a distinguished professor of cognitive psychology at the University of California in San Diego, has studied consciousness for decades. He explained, "The question of what is consciousness is something that we've thought about as human beings for millennia." However, the problem arises when we try to explain the occurrence of consciousness through "physicalist" approaches, meaning physical properties in the brain. For him, intuitively, "the problem of consciousness" can be appreciated at a very basic level through our personal experiences, even everyday ones such as "the feeling of a headache, the smell of garlic, what it is like to experience the feeling of love, or to smell chocolate, or hear the sound of a trumpet," and "being aware." He agreed with Sestan's view that science has struggled to explain even something as simple as what it is "like to taste chocolate."

David Chalmers, an Australian philosopher at New York University, has summarized the problem very well. He says, "Consciousness poses the most baffling problems in the science of the mind. There is nothing that we know more intimately than conscious experience, but there is nothing that is harder to explain." Chalmers calls this the "hard problem" of consciousness. This contrasts with the "easy problems," which essentially involve understanding the mechanisms that allow the brain to deal with the various sets of information that it receives.

Dr. Megan Craig further clarified Chalmers's point. She explained that contemporary debates around consciousness are often talked about in terms of the "hard" and "easy" problems of consciousness. However, "this may be misleading because they probably are all hard problems. But what is meant by the hard problem of consciousness is the relationship between the physical brain and consciousness." She explained this modern debate actually "dates right back to this ancient problem of psyche and body or mind and body and how to bring the two together." In that sense, then, "The easy problems of consciousness would be those problems that are essentially solvable through neuroscience and through a more and more sophisticated and minute investigation of the structure and the mapping of the brain." The "easy" problems involve "things like trying to determine what happens to the brain when it is stimulated in certain ways and localization of activity in the brain"; whereas the "hard" problem "is really hard because you are trying to establish the relationship between the physical entity [the brain] and this [seemingly] non-physical entity, [which is] consciousness." The hard problem seems unsolvable because "consciousness itself is not able to be located in the same way as a given stimulus and response can be located in the brain." She was referring to the fact that modern medicine has answered some of the relatively "easy" questions regarding the relationship between thoughts, feelings, and the brain—namely, the specific areas that are involved with certain feelings, emotions, and thoughts (through mapping techniques)—but not how thoughts can be produced from brain cells. That is the "hard" problem.

Hoffman explains that the "problem" arises when people "try to assume the brain is responsible for producing consciousness." That is because even though most people, including nonspecialist scientists, may think brain processes create our mental world, the reality and hence the "problem of consciousness" is the absolute lack of evidence to support this. Science is driven by experimentation and data. Yet, as Hoffman puts it, there is still not a single piece of data or any experimentation to show how brain processes may give rise to a single thought, not even the simple experience of the taste of chocolate, let alone the complex and unique mental universe that makes us all who we are and the Einsteins, Beethovens, and Mozarts of this world who they are.

Hoffman points out, scientists are those who say "it must somehow happen, even though there isn't any evidence to support this view." Other scientists point out that no one has ever put forward a plausible biological mechanism to explain how this can occur, let alone any actual scientific studies. As Dr. Seth's article highlighted, with some of those so-called reductionist views, people just believe that at a certain level of complexity of brain cell connections, consciousness and the complexity of the human mind just happen, again almost as if by magic.

We still have so much to learn. I liken the search for consciousness to the search for weapons of mass destruction in Iraq. Many people were certain that there were weapons of mass destruction before the mass invasion in 2003. They showed pictures with satellite images of buildings and warehouses. Aside from real intelligence experts, so many people more or less believed them. However, no weapons were ever discovered, even after troops arrived on the ground. Similarly, some scientists often show pictures and scans of the brain, but that doesn't show the full story—or that you're going to find the mind. You may fully understand the structures of the brain, but that doesn't mean you can solve the mind–body problem. Nor does it explain the origin of self.

All of this is hugely relevant to the new, emerging field of study beyond the boundaries of death. Think back to the studies in reviving pig brains after death, and to when this technology is inevitably adapted and used to help revive humans after death as we know it today. Do we assume consciousness is produced by the brain and so dies? Or do we take seriously the emerging evidence from studies like AWARE-I and -II and the experience of many millions of patients who have returned from the grey zone of death? If we don't understand how our consciousness comes into being, how can we understand consciousness beyond the threshold of death? Many of our respondents noted feeling a sense of vast expansion of their consciousness and a sense of separation and liberation from their bodies. They reported feeling very clearly that they were not their bodies or a sense of relief at shedding their bodies like a heavy and burdensome winter coat. Yet they still retained a sense of themselves as individuals, as the person who went by their name and had lived their life experiences. And most crucially, our interview subjects survived their trip to the grey zone and back. Their consciousness felt like it had been stuffed back into a reduced state as it "returned" back to their body, and once again appears to be firmly rooted within it. They diligently answered our questions but could not speculate at how their consciousness and body separated and rejoined.

The definitive direct answers to these questions will inevitably come when science finally discovers the nature of human consciousness. However, for now, we have to evaluate all the data that is emerging through science in an unbiased manner to help us understand the nature of human consciousness in life and in death.

Chapter 14

THE SELF AND ITS BRAIN

ECCLES, DESPITE HIS UNDOUBTED ACCOMPLISHMENTS IN NEUROscience, did not agree with the direction and theories that many of his colleagues were putting forward to explain consciousness and that people like Dr. Seth have now followed. Eccles argued in *The Self and Its Brain* that the mind and consciousness (the part of an individual that experiences, observes, and is aware of life) is something separate from the body; the mind and not the machinery of the brain provided the unity of conscious experience. By that he was referring to the fact in your conscious mind, you have a single experience of yourself, rather than millions or billions of little "me"s, which is what you would expect if your selfhood was produced by your brain cells. Further, he thought that the mind actively selected and integrated brain cell activity and moulded it into a unified whole. Just as Megan Craig had experienced firsthand, Eccles considered it a mistake to think that the brain did everything and that conscious experiences were simply a reflection of brain activities, which he described as a common philosophical view. "If that were so," Eccles said, "our conscious selves would be no more than passive spectators of the performances carried out by the neuronal machinery of the brain. Our beliefs that we can really make decisions and that we have some control over our actions would be nothing but illusions."

Sir John's theory was that the brain, in particular an area toward the front of the brain called the supplementary motor cortex, acts as a liaison through which the mind interacts with the brain. *Yet the mind does not reside within it.* That is to say, no one knows where the mind is, or from where our conscious self emerges, and no one since his remarkable discoveries has been able to give a clear answer.

In essence, Eccles argued that who we are—our mind, our selfhood, consciousness, and awareness—is not the brain but is intricately linked to the brain, and in turn modulates the brain. Scientists, of course, can find signs of this in the form of brain processes, which can be thought of as a sort of "signature" of the mind and consciousness in the brain. However, for Eccles and many other scientists, these electrical and chemical processes—*the neural correlates of consciousness*—represent the "tethering" effect of the mind on the brain, just as Megan Craig had experienced firsthand. In other words, activities of our mind—our thoughts, emotions, and all our inner states—are not produced by the brain. Instead, they independently manipulate brain processes, and it is this interaction between a mind and consciousness that is separate from but attached to the brain that scientists detect. These "signatures" of the mind may be in the form of electrical waves with different frequencies or chemical surges in dopamine, prolactin, and oxytocin across different parts of the brain when we experience different emotions, such as love.

Of course, being in a tethered state means that although the mind modulates the brain, the brain, too, modulates the mind. The relationship is not one way. It works both ways. *The key is that while brain processes affect our mind and consciousness, they don't produce them.*

We can use an analogy of a hand in a glove to explain this better. When we place our hand in a glove, the two become intricately linked together. Yet they are inherently two different substrates. Imagine yourself wearing a glove. Of course, the shapes of your hand and glove look alike, and they intimately follow every movement together. If, one day, the glove is too tight, it will impact your hand, as it will create a sense of discomfort. At the same time, the movement of your hand also impacts and modulates the movements of the glove. These two modulate and interact with each other. In fact, the two are so intricately tethered to each other that to an outsider, they may appear the same. Yet they are not. How the "hand" and "glove" separate is a key idea in understanding the experience of lucid hyperconsciousness in the grey zone.

MANY OTHER SCIENTISTS BESIDES ECCLES HAVE ARGUED THAT brain-based theories cannot fully explain the observed features of consciousness. As Dr. Hoffman earlier explained, "physicalist" approaches do not seem to provide a satisfactory answer. However, these views are often ignored by those who don't agree.

Stuart Hameroff and Roger Penrose have grouped the limitations of the conventional or "physicalist" (reductionist) theories in four broad categories.

The first and most obvious limitation is that they do not provide a plausible mechanism to account for the development of consciousness or thoughts from brain cell activity. The theories propose potential intermediary pathways but still do not answer the fundamental question of how brain cell activity might create thoughts and consciousness or human experience. Remember the man who was missing 90 percent of his brain but still had normal consciousness? Brain cells, like any other cell, can manufacture protein-based substances. They can even be linked with electricity, but the nature and substance of thought seems inherently different from electricity or a chemical or any protein-based substance in the brain, as Eccles highlighted.

Second, another limitation relates to how activities distributed all over the brain—through the actions of billions of individual brain cells at any one time—can eventually bind into a single unitary sense of self that leads to the notion of "I" and our selfhood. This was also highlighted by Eccles. Remember, we see ourselves as one person, not billions and billions of tiny selves, each linked to one of our billions of brain cells. I see myself as one person, not a fractured amalgamation of infinite tiny selves. Probably you do, too.

Third, how do chemical or electrical events that are continuously taking place in different parts of the brain—such as the release of various hormones—but are not part of our "consciousness" suddenly become conscious, other than to say that it somehow does occur at a critical point?

Fourth, and perhaps most important of all, we know that a fundamental part of our lives involves the notion of free will and our choices. We are judged in society based upon our intentions, actions, and choices. Yet the brain-based views cannot account for this reality. If correct, they would mean that our lives are completely determined by our genes and environment, and hence there would be no place for personal accountability.

Several recent philosophers and writers have suggested that free will is an illusion. However, most people reject this theory based on everyday experiences. The theories that our actions are solely due to our genes acting in combination with their environment don't hold water. If they were true, no one would be held accountable for anything, and equally, nobody could be rewarded or acknowledged for anything.

In this theory, Nenad Sestan's accomplishments, his entire journey and decisions, would have been initiated by the simple mechanics of his brain rather than any intrinsic, personal motivation. The same would apply to others, such as Einstein, Beethoven, Mozart, and so on. The actions of these geniuses are not choices, according to this theory.

In the absence of personal choice and accountability, should the Nobel Prize Committee and other committees that strive to acknowledge human accomplishments stop awarding prizes? And should all the Nobel laureates return their prizes?

What about people who show courage and put their own lives at risk for a comrade on the battlefield? Should society not recognize their valor?

What about all the people who strive to achieve academic excellence? Is it just their brain that automatically studies? Clearly not, because, as any student knows, they have a choice, and they must force themselves to study through willpower, as opposed to blowing off classes and having fun with friends.

Any attribute by this thinking, good or bad, is simply the brain reacting to genetics and environment, and not due to our personality, dreams, hopes, and sense of social obligation or personal ambition. Accordingly, the good do not deserve praise, and the bad do not deserve punishment.

However, as Sir John Eccles pointed out, we are not just passive spectators of what the neuronal machinery of our brain does. We choose from a number of options, and society holds us responsible for our choices. Our experience that we make decisions and that we have some control is not an illusion. It is very real. This is what the fabric of society is based on.

This limitation of all the brain-based theories has led some to suggest consciousness may in fact be an irreducible scientific entity in its own right, similar to many of the concepts in physics, such as mass and gravity, which are also irreducible entities. Other discoveries, such as the discovery of electromagnetic phenomena in the nineteenth century or quantum mechanics in the twentieth century, were also inexplicable in terms of previously known principles.

Throughout his own lifetime, Eccles's position had been that there was "a combination of two things or entities: our brains on the one hand and our conscious selves on the other." He thought of the brain as an "instrument that provides the conscious self or person with the lines of communication from and to the external world." It does this "by receiving information through the immense sensory system of the millions of nerve fibres that fire impulses to the brain, where it is processed into coded patterns of information that we read out from moment to moment in deriving all our experiences—our perceptions, thoughts, ideas and memories."

According to Eccles, "*We as experiencing persons do not slavishly accept all that is provided for us by our instrument, the neuronal machine of our sensory system and the brain, we select from all that is given according to interest and attention and we modify the actions of the brain, through 'the self' for example, by initiating some willed movement.*" In saying this, Eccles was making an important point. The self independently analyzes the information provided to it through the apparatus of the brain and selects what to do with that information. We have the power to select from options rather than be what the brain automatically does.

Dr. Donald Hoffman supports a position similar overall to that of Eccles. He explained that although he, too, had started out like maybe 95 percent of his physicalist (reductionist) colleagues, who say that "it is the functioning of the brain that creates the mind and consciousness" and that "the mind is what the brain does," he no longer agrees with that view. He said, "I was impressed with the failure of the physicalist approach to solve this hard problem of consciousness." His idea was "to come up with a scientific theory in which consciousness is fundamental."

Dr. Maxwell Bennett is an eminent professor of neuroscience from the University of Sydney and scientific director of the Brain and Mind Research Institute. He, like Eccles, also makes a distinction between brain processes and the choices that we humans make. He pondered about violence. When a person throws a punch, he asks, did their malfunctioning brain cause the punch to be thrown? Should they be punished or understood? Bennett is one of the most qualified people in the world to investigate the tremendous complexity behind this question, and his conclusion, like that of Eccles, is clear. He says, "We cannot attribute to a part of the brain that which we human beings do: the thinking, deciding, admiring, singing. We human beings do these things, not the brain."

Dr. Bahram Elahi, who has put forth the hand in a glove analogy,* is a professor of surgery and anatomy with a strong interest in human consciousness and the self. In his works, he has expressed the view that although the brain and consciousness are separate, what gives rise to consciousness it is not immaterial either. Rather, it is composed of a very subtle type of matter that, although still undiscovered, is similar in concept to electromagnetic waves, which are capable of carrying sound and pictures and are governed by precise laws, axioms, and theorems.

Therefore, in Bahram Elahi's view, everything to do with this entity should be regarded as a separate, undiscovered scientific discipline and studied in the same objective manner as other scientific disciplines. He argues that as science is a systematic and experimental method of obtaining knowledge of a given domain of reality, then human "consciousness," or the self, can and should also be studied with the same objectivity.

---

* See chapter 13, *Fundamentals of the Process of Spiritual Perfection: A Practical Guide* by Bahram Elahi (New York: Monkfish Book Publishing Company, 2022).

Each scientific discipline, such as chemistry, biology, and physics, has its own laws, theorems, and axioms, and in the same manner, the human soul or self, which is consciousness, should also be studied in the context of its own laws, theorems, and axioms.

In his view, human consciousness, the self, is also a scientific entity and a type of matter, though it is a substance that is too subtle to be measured using the scientific tools available today. Therefore, the brain is an instrument that relays information to and from both the internal and external worlds, but "consciousness" is a separate and subtle scientific entity that interacts directly with it. This entity and its functions and structure may also hold the key to understanding the recalled experience of death and uncovering the mechanisms that lead to the experiences outlined in the previous chapters. Though it is mysterious now, eventually we will have some tangible insight into how it works and what happens when an individual wades into the grey zone of death.

As a philosopher, Megan Craig understands and has personally experienced the critically important point that Eccles and many others, whose views we have reviewed and who do not support the physicalist view, have made. She recognizes that humans have been asking the question of what consciousness is for millennia, and that it remains scientifically and philosophically unanswered. Of course, in the absence of scientific data and evidence, people, whether scientists or otherwise, naturally gravitate to one or another philosophical view, usually one that matches their own prior personal views, which may not necessarily represent the truth of the matter.

As evidenced by Anil Seth's article, even the views of Sir John Eccles, a Nobel Prize–winning brain scientist, are typically omitted and ignored by some in the contemporary scientific community. So, too, are the views of many other scientists who do not agree with physicalist (reductionist) views. This active and purposeful omission, almost like a type of censorship, gives rise to a false and biased impression, especially in the media, that all major scientists support one overall view: that brain processes produce consciousness. Worse still, it gives the impression that there may be some evidence to support this view. After all, scientists are supposed to form their opinions based on scientific data. Otherwise, what makes them stand out compared with others?

The theories selected by Anil Seth had one thing in common: they were all reductionist views. Yet without any data or evidence to support them—to Eccles's point—we are dealing with opinions, beliefs, and philosophy, not science. Data is what distinguishes science from philosophy and has led to success in understanding the world around and beyond us. Although interesting, all the so-called physicalist theories represent a continuation of what some philosophers had broadly argued for thousands of years.

To better understand the philosophical roots of today's discussions, we need to go back in time. Greek philosophers at the time of Plato and Aristotle referred to what modern scientists broadly call consciousness today as the *psyche*. For those Greek philosophers, the psyche had a very specific meaning. It was used to refer to what animates and gives living beings, notably humans, their unique characteristics, including what shapes their identity and leads to their sense of morality. In short, the psyche was considered the distinguishing mark of all living things, not just humans. However, in humans, the psyche also included emotional states and mental and psychological functions, such as thought, perception, desire, planning, and practical thinking, as well as moral qualities and virtues. Just as we see now, back then, too, there were many different and conflicting theories about the nature of the human psyche, or consciousness, and its relationship with the body, as well as what happens to it after death.

Some may wonder about the relationship between the ancient philosophical concept of the psyche with the soul and the more modern term *consciousness*. In reality, these three terms all represent the same thing. The Greeks used the term *psyche* first, and this was later translated into the Gothic term *sêula*, meaning "life, spirit, consciousness." From there, it was translated into the Old English term *sáwol*. This is where our more familiar term *soul* comes from. Contemporary scientists, hoping to avoid reference to religious traditions, increasingly use the term *consciousness.* Most people associate "psyche" with the mind and the "soul" with something vague, esoteric, and religious. In reality, along with "consciousness," they all refer to the same thing: that unique inner universe of consciousness that comprises the "self."

Ernest Arbman, a twentieth-century Swedish scholar, suggests that even in ancient times, the psyche, or soul, was seen as an intrinsic part of human psychological functions, unlike the way many people view the soul today. By the beginning of the classical period, in the fifth century BC, the Greek world was dominated by the work of natural philosophers. These innovative thinkers hoped to explain what all matter is made of, breaking it down to its most basic materials. They were also among the first people to try to understand the natural world without the lens of supernatural or mystical explanations. They enthusiastically embraced the idea of rational and critical discussion. Among the topics they discussed was the nature of the psyche, or soul.

Those early Greek philosophers disputed many questions related to the soul, or the psyche. This included understanding what animates living beings as well as the source of human thoughts, the mechanism of cognitive activity, and the nature of emotions, perceptions, and voluntary movement. They also debated other issues concerning the essence of the psyche, or soul, including the location of the intellect and even the causes of neurological and psychiatric disorders. There was also great discussion regarding the relationship of the soul with the body and its overall outcome after death.

Contrary to many people's perceptions today, the early definitions of the psyche, or soul, among the Greek philosophers did not always include belief in an immortal psyche, or soul, or even an immaterial psyche, or soul. In fact, there was fierce debate on this subject, as there is today. Pythagoras (fl. 530 BCE) believed that the soul was of divine origin and existed both before and after death, while Democritus (460–370 BC), who is credited with atom theory, as well as Epicurus (341–271 BC), a supporter of this theory, believed that the soul, like the body, was made up of small indivisible parts called "atoms" and therefore, after death, these "soul atoms" dispersed, and thus nothing remained of the soul.

However, as Dr. Megan Craig pointed out, among all those philosophers, it was Plato's and Aristotle's views that have most shaped our thoughts on the subject. Plato believed that the physical world that is perceptible through our five senses, although "real," is less real and less "perfect" than another domain that is imperceptible with our senses and belongs to the realm of the psyche, or soul. Thus, he believed the soul, or psyche, was an immaterial substance.

Megan Craig explained, "Platonic thinking about the psyche is deeply rooted in the idea that the psyche predates the body. The psyche is initially tied to something divine and at a certain point, the psyche descends or falls into a form, and this might be a human form. So, the physical body in Platonic theory is really described as a shell or even a prison that encapsulates the psyche just until that moment when the psyche can be released again and return to the path of the divine or return to the immortal realm. This is really an account of the body as something temporary and transient that is impinging on the psyche's ability to return to its place of origin."

By contrast, Aristotle, a student of Plato, believed that the matter and the form of a being cannot be separated, but can be distinguished from one another. He believed the form of any being arises from the characteristics of its matter, rather than existing in another immaterial realm, as Plato did. He theorized that the psyche, or soul, in humans is actually a by-product of the activity of the physical matter.

For Aristotle, then, the soul is to the body what vision is to the eye. If the eye works perfectly, then as a by-product, you have vision. However, vision isn't the same as the eye; it is the "soul" of the eye. Further, the soul, or psyche, in humans—their thoughts, feelings, and psychological makeup—is a by-product of the perfection of physical matter. For as long as the body works, you have a soul, or psyche, and when the body ceases to function, you no longer have it. He believed that since the soul is produced by the body, it therefore cannot exist without it and perishes after death, aside from one aspect of the psyche—the intellect—which could remain after death.

Although myriad opinions were expressed by other philosophers and scholars in the years that followed, the contrasting and distinguishing views of these two classical Greek philosophers have gone on to significantly shape the views of many later Western as well as Islamic Golden Age philosophers and, from there on, those of our modern scientists. As Dr. Maxwell Bennett writes in his article "Development of the Concept of the Mind," it was these two broad views that "were held by competing scholars and theologians during the next 2000 years."

THE ROOTS OF THE PHILOSOPHICAL VIEW THAT THE BRAIN PRODUCES mind and consciousness can also be traced back many centuries. By the fifth century BC, two main theories, *encephalocentrism* and *cardiocentrism*, had also become widely established regarding the origin of human thoughts. The former considered the brain to be the seat of human consciousness, sensation, and knowledge, whereas the latter attributed these qualities to the heart.

Both the encephalocentric and cardiocentric theories continued to be disputed until at least the Renaissance, following which encephalocentrism became the predominant worldview. Two Italian scientists, Crivellato and Ribatti, wrote that "it is clear that many leading concepts in the modern world including those within neuroscience and psychology find their origin in the speculation of ancient Greek scholars."

Today's scientists, unlike the ancient Greeks, agree on the central importance of the brain, but their views on the subject of the psyche, the soul, or consciousness can be more or less divided into two broad categories. There are those who support a view that broadly corresponds with Aristotle's view, as Dr. Donald Hoffman explained, that "the mind is what the brain does," and those who broadly support Plato's view that the mind and consciousness are separate from the body.

Francis Crick, the codiscoverer of DNA and another Nobel Prize winner, was a notable proponent of the reductionist belief that the "self," or soul, arises from the brain and that soul is extinguished when the body dies. Sir John Eccles was the most prominent scientist in the other, non-reductionist group, which considers the essential reality of the human psyche, or soul, to be a separate entity from the brain and body and believes that when you die, the psyche, or soul, continues as a different type of matter. However, the jury is still out.

Dr. Megan Craig, like Eccles, Hoffman, Elahi, and many others, falls in the non-reductionist group. She has said that she just doesn't think that it's possible to answer this question by looking more intently at the matter of the brain. She said that "it is in part because of my own experience and having the experience of feeling really adjacent to my brain, like a witness." This sense of witnessing our brain as something different from ourselves is, of course, something that we can all relate to. It does not require us to undergo a concussion. Like Megan, in practice we can all distinguish between ourselves and our body apparatus in our everyday life experience. For example, we say "my hand," "my arm," "my heart," and even "my brain." We say this because, in reality, we experience the components of our body, including our brains, *as belonging to, but distinct from, our self.*

MEGAN'S DESCRIPTION OF HER FIRST-PERSON EXPERIENCE, WHICH enabled her to witness her selfhood and sense of awareness as being different from her brain, is reminiscent of the famous "floating man" (or "flying man") thought experiment conducted by Persian polymath and physician Avicenna (980–1037 CE).

Avicenna, who is regarded by many as the father of early modern medicine, is one of the most influential philosophers and writers of the Islamic Golden Age. Insatiably curious, he wrote nearly 450 works in total in his lifetime, of which it is thought that about 240 remain. Among his most famous works are *The Book of Healing*, *The Canon of Medicine*, and a medical encyclopedia; these were the standard medical books used in many medieval European universities well into the seventeenth century.

During a stint in prison, when he was locked away in the bowels of a Persian castle, Avicenna wrote what has since become known as his famous "floating man," or "flying man," thought experiment. He asserted: suppose a man is suddenly brought into existence out of thin air, in a stroke—as a fully developed, fully grown, and mature human being—but who is fully suspended in thin air. In this suspended state, he will be without any prior existence, and so will have no memories. Crucially, in this state, in midair, he will be suspended with his limbs splayed, but he is also not in contact with his body, and all five of his senses are disabled. There will be no wind to feel and no sensory input from any of his other senses either. So there is nothing for him to experience: nothing to see, hear, smell, taste, or touch. In this state of radical sensory deprivation, without any prior memories of any existence, he can't form any new experience either.

The question Avicenna then poses is: What, if anything, would this flying man be aware of? Without any prior experiences to draw upon, Avicenna claims that in this state, the man would still be aware of one thing, and that is his own existence—*his selfhood*. There is no doubt that he would affirm his own existence, while not affirming the reality of any of his limbs or inner organs, his bowels, or heart or brain, or any external thing. And if it were possible for him in such a state to imagine a hand or any other organ, he would not imagine it to be a part of himself or a condition of his existence.

Self-awareness is fundamental to our mental life. Given that even without being aware of the existence of his body—and without any sensory input or any prior experience, which is what shapes humans—he would still fundamentally be aware of himself, and this suggests that the self, psyche, or consciousness is fundamentally independent from bodily processes and sensory physical experience.

Centuries later, the seventeenth-century French philosopher Descartes stated, "I think therefore I am." Both philosophers were expressing that even if we doubt our senses, we cannot doubt our existence as a being who experiences the world around us. Likewise, we are not simply a straightforward physical thing.

One of the people whose works Megan teaches a lot is the famous nineteenth-century American psychologist and philosopher William James, who wrote, "Experience overflows. There's something necessarily uncontainable about experience and as soon as you try to focus on it, or fix it, or locate it, or pin it down, it's already moved beyond the frame that you've tried to impose on it." Megan explains, "That is the way that I think about consciousness in the brain. Not that the brain is not an important and critical feature of what it is to be alive . . . but I do not think that even the brain in all of its intricate complexity is a wide enough, or malleable enough frame for consciousness and what we call experience."

The problem becomes clearer if we, like Megan Craig, try to ponder all the events and phenomena that occur in our mental universe. Then, like Sir John, we will probably have to ask ourselves: Are we simply the sum of electrical impulses or chemical signals that crisscross one region of the brain to another? In short, in the same way that the electrical circuitry in a computer doesn't explain the content within the computer, which is generated by the people using the computer, so the functioning of the electrical grid or circuit in the brain, as Eccles believed, doesn't explain the "content" that the brain expresses.

Since Sir John passed away, in 1997, we have seen many advances in neuroscience thanks in part to his research. But we still don't know why we humans—alone in the animal kingdom—can paint works of art, write poetry, create movies, read books, cure diseases, contemplate ethics and morality, and even solve mathematical equations that can help launch us to Mars.

Now, decades later, the importance of trying to find answers to Eccles's questions in an open and unbiased manner is ever more critical. The discoveries that we have discussed over the course of the book—gamma waves in dying brains, the pigs head study that showed the grey zone of death is a wider continuum than previous acknowledged, and, of course, the AWARE studies that detail the experiences of survivors—demand a new way of thinking about the self and consciousness. It is likely that in the near future we will be able to reverse death and resuscitate people hours later than ever before.

Where is the line between life and death? And if there is no clear true biological point that can distinguish life from death, will we start to consider life and death in the wider and more meaningful sense of what makes us all who we are as conscious, thinking beings, as entities with consciousness? Isn't that what life really refers to—being conscious? This reflects the essence of Eccles's questions, when he, too, pondered how is it that we think, have personalities and feelings, and retain memory. How do these activities that mark the self relate to the brain? What happens to that entity as we transition from life through to death? And how is this different than, say, other experiences that often become inaccurately conflated with the recalled experience of death?

Conventional science and medical practices are still a long way from accepting the continuation of consciousness beyond the traditional boundaries of death as a fact. In a way I'm sympathetic. If we accept the recalled experience of death, in the way that is being described by those many millions of survivors, and the way that scientific data is beginning to support their claims, then essentially everything we know about treating patients at the end of life and long-standing views on the relationship between consciousness and the brain will change. And change doesn't come easy to any of us. This is why many people have tried very hard to reject these claims, and they have even come up with some ludicrous ideas and explanations. Still, it's important to push back at the other theories that are erroneously used to explain the recalled experience of death and demonstrate why they are not an appropriate or accurate explanation for the phenomenon.

Section Four

A WORLD OF DISTORTIONS

Chapter 15

PUTTING LIPSTICK ON A PIG

In the spring of 1916, three exhausted sailors stumbled toward a whaling station on the north coast of South Georgia, near the continent of Antarctica. Starving, frozen, and exhausted, the men had been walking for the better part of thirty-six hours in extremely cold and dangerous conditions, climbing over icy mountains with nothing more than a rope and an axe and no provisions. These three, Ernest Shackleton, Frank Worsley, and Tom Crean, were ultimately able to secure help for the rest of their surviving sailors of the *Endurance*, iced in and stuck on the south side of Elephant Island.

In 2015, *The Guardian* newspaper described how the three explorers "did not talk about it at the time, but weeks later all three men reported an uncanny experience during their trek: a feeling that 'often there were four, not three' men on their journey. The 'fourth' that accompanied them had the silent presence of a real person, someone walking with them by their side, as far as the whaling station but no further. Shackleton was apparently deeply affected by the experience, but would say little about it in subsequent years, considering it something 'which can never be spoken of.'"

In his book *South: The* Endurance *Expedition*, Shackleton himself explained, "I know that during that long and racking march of thirty-six hours over the unnamed mountains and glaciers of South Georgia it seemed to me often that we were four, not three. I said nothing to my companions on the point, but afterwards Worsley said to me, 'Boss, I had a curious feeling on the march that there was another person with us.' Crean confessed to the same idea."

These encounters are known as "third man" encounters, after the T. S. Eliot poem *The Waste Land*:

Who is the third who walks always beside you?
When I count, there are only you and I together
But when I look ahead up the white road
There is always another one walking beside you.

In the book *The Third Man Factor*, John Geiger details several similar stories from mountaineers, sailors, and survivors of terrorist attacks over the centuries. All those people that he investigated recalled similar experiences—a strong impression of a vague presence—usually without a clear form.

Writing in the same article in *The Guardian* newspaper, researchers Ben Alderson-Day and David Smailes speculate these "third man" encounters may result from severe trauma or other extreme situations (as in the case of Shackleton), such as social isolation or some great loss (bereavement). They also say it might even be "tempting to think" that the brain has a "hallucinatory defense mechanism" to shield itself from feeling completely alone, abandoned, or lost. To support their argument, they point to the fact that "following bereavement, for example, many people report sensing the presence of their deceased loved one; a feeling that someone is still in the house, just upstairs, or in their favorite chair."

Likewise, they say, people report "presences" during sleep paralysis, when an individual feels awake but is unable to move their body. Often, they say, this is accompanied by "physical sensations such as pressure on the chest and difficulty breathing." Importantly, the feeling of a presence is also reported in some brain disorders, such as Parkinson's disease, and after damage to specific areas of the brain.

To these authors, the different contexts in which the "presence" occurs may provide clues about what might be happening in the brain. They argue that these emotional factors, such as bereavement and an expectation that someone "should" be there, are important elements in these experiences.

What this article also highlights is that when faced with something complex and unknown regarding human experiences, some scientists try to oversimplify things. In this case, the researchers were assuming the explanation for any feeling of a presence is more or less the same, even though the conditions they were pointing to are all so different.

They go on to also suggest these experiences are likely to occur in people suffering from Parkinson's disease and on high doses of medications, which could suggest the neurotransmitter dopamine is involved. They further speculate that damage to a brain area "called the temporoparietal junction (TPJ), and electrical stimulation of this area can induce the feeling of a nearby sense presence."

The TPJ is a region located halfway across and to the side of our brains that contains within it the *angular gyrus*. Some scientists believe that it helps maintain a clear representation of what our body looks like for ourselves. That is why Alderson-Day and Smailes suggest a seizure or other ailment in that location may cause the reported "feelings of a presence." *They propose that a disorder to this part of the brain causes the inner representation, which we all have of our own body, to somehow become "duplicated or projected outside of us (known as an 'autoscopic' experience)."*

People have known about autoscopic experiences for millennia. In fact, the term *autoscopy* is derived from the Greek words *auto*, meaning "self," and *skopeo*, which means "looking at." Although many believe it was the Greek philosopher Aristotle who first described it, it is now thought that Dostoevsky popularized it through his nineteenth-century novel *The Double*.

These two researchers used the term when talking about the experience of a presence; however, typically this term has been used to refer to a phenomenon that is much more elaborate and intriguing. In a pivotal article published in the *British Journal of Psychiatry* in 1994, Drs. Dening and Berrios, from the Department of Psychiatry at the University of

Cambridge in England, explained that autoscopy is "*a visual experience where the subject sees an image of him/herself in external space, viewed from within his/her own physical body.*"

In essence, *people experience seeing a physical double—a replica of themselves—at a distance, which can sometimes even be acting independently and separately.* The key point is that this virtual self is viewed from within one's own body. The replica (or double) may be seen standing across the room and cooking, reading, or performing other activities, while the person remains fully aware that they are sitting on a chair or lying in bed. *Throughout autoscopy the person realizes that the duplicate is not themselves and is in fact unreal, but they struggle to rid themselves of it, sometimes with potentially devastating consequences.*

In Anil Ananthaswamy's book *The Man Who Wasn't There: Tales from the Edge of the Self*, he describes an extreme case of autoscopy. He recounts in the autumn of 2011, he met Peter Brugger—then a PhD student at the University Hospital of Zurich—who told him of a twenty-one-year-old man with epilepsy who had apparently come "face-to-face" with his own double and had unintentionally almost killed himself while attempting to rid himself of it.

The incident happened when the young man had decided to stop taking his antiseizure (epilepsy) medications. Ananthaswamy wrote, "One morning, instead of going to work, he drank copious amounts of beer and stayed in bed. But it turned out to be a harrowing lie-in. He felt dizzy, stood up, turned around, and saw himself also lying in bed." Seeing a double was disconcerting. At one point, he jumped on his duplicate virtual self and tried to shake him awake. When that wouldn't work, he felt so disconcerted that he couldn't take it anymore. In the end, he decided to jump out of a four-story window in his attempts to free himself of it. He wasn't trying to commit suicide; he simply couldn't take the sensation of seeing a virtual second version of himself. The man survived, but while being treated for his injuries, his doctors discovered a brain tumor on his left temporal lobe. Once the tumor was removed, he never had the autoscopic experience again.

In a medical article, "'Seeing Oneself': A Case of Autoscopy," Drs. Giovanna Zamboni, Carla Budriesi, and Paolo Nichelli, from the Department of Neuroscience at the University of Modena, in Italy, described a young woman who, after suffering brain injury, had recounted to her physician that she had suddenly started to see her own duplicate virtual image. Just like a mirror image, her double replicated her movements. For instance, any facial or arm movement she executed was reproduced in real time by the image. Wherever she looked, she saw her mirror image right in front of her, at a distance that varied depending on where she was looking. The image of *her duplicate, virtual self* appeared about a meter away if she was looking into the distance. It seemed to be lying on the floor if she was looking down, and it appeared on the ceiling when she was looking up. In short, it would appear in front of her line of sight whichever direction she looked in.

When she was being examined by her doctors, she found her duplicate interposed between herself and the doctor. If a solid object (for example, a sheet of paper) was placed between the mirror image and herself, she still saw the image, but nearer to her, on the surface of the paper. It was as though the image, *which was transparent*, was set "on a sheet of glass" resting against whatever object she was looking at. The image was somewhat blurred, and she could not make out its colors or small details. It disturbed her vision of other objects, even though, *being transparent*, she could still see through it.

In a separate scientific article published in 2011, "Autoscopic Phenomena: Case Reports and Review of Literature," Dr. Francesca Anzellotti and her colleagues from D'Annunzio University in Italy described another case. This time, of a forty-year-old teacher who had presented to their hospital with seizures and had been hooked up to a brain electrical wave monitoring—electroencephalography (EEG)—machine. Before the onset of one of her seizures, the woman reported an unclear change in her sense of awareness of her body. She had then signaled by a hand gesture *the abrupt appearance of her entire body in front of her—in an upright position, motionless, silent, and of normal size, with her same clothing and facial expressions.*

This sensation coincided with her seizure, as evidenced by abnormal brain spikes on the EEG brain monitoring machine during her experience.

Then, as her brain waves slowed back to normal and her seizure ended, the woman signaled to her doctors that the image had also disappeared. So, like with the man who had jumped out of the window, this woman's autoscopy occurred during a seizure and disappeared when the seizure ended.

All of these examples are a form of visual illusion—a trick of an abnormally functioning brain—that causes people to hallucinate seeing an unreal virtual double of themselves at a distance. Interestingly, sometimes, instead of seeing a whole virtual self, people may only see a portion of their body, say just the arms or legs or trunk.

In his book, Anil Ananthaswamy describes a conversation with Dr. Olaf Blanke, a neurologist at the Swiss Federal Institute of Technology and founding director of the Center for Neuroprosthetics in Lausanne, Switzerland. In recent years, much of Blanke's work has focused on exploring how technology—*in particular, robotics and virtual reality*—can be adapted to help people suffering with phantom limb syndrome.

As its name suggests, phantom limb is a condition in which amputees sense a ghostly version of their amputated limb is still present. This is sometimes excruciatingly painful. Olaf Blanke explained, "A sensed presence is like experiencing a full-body phantom: if a phantom limb is the continued sensation of having a limb that has been amputated, then a sensed presence of a body is its full-body analogue."

Throughout his career, Blanke thought the experience of separation reported by people in death is a similar illusion. In September 2002, a headline on the US-based broadcaster CBS ran "Out of Body Experiences Explained." This was one of many headlines that had appeared across the global media outlets following the publication in the leading scientific journal *Nature* by Dr. Blanke of a bizarre occurrence in a forty-three-year-old woman who was being investigated for intractable seizures.

Dr. Blanke and his colleagues had implanted electrodes to stimulate different portions of the woman's brain to help identify the abnormal area responsible for discharging electrical seizures. While undergoing stimulation of her brain, quite unexpectedly the woman reported experiencing a different perception of herself. When *the right angular gyrus of the temporoparietal junction of her brain* had been stimulated, she felt like she was "sinking into the bed." With further stimulation, she felt like she was "falling from a height," before seeing a virtual image of her own legs and lower half of her body, as if from above. Specifically, she stated: "I only see my legs and lower trunk."

However, she never described a sensation of separating from herself or being able to see real events occurring in the room, whether the medical procedure or anything else. She said her legs were "becoming shorter," and then "appeared to be moving quickly towards her face." This was frightening and disconcerting enough that she even took evasive action because she thought she was about to be hit in the face by her own legs!

Based on these illusory *distortions of body image*, Olaf Blanke did something quite uncustomary. He arbitrarily labeled this woman's experience an "out of body experience," *even though she had never claimed to have experienced being out of her body.*

The relabeling of visual illusions as an "out of body experience" was quite a cunning move because if Blanke had labeled them as "visual illusions," then they would not have seemed extraordinary or newsworthy. Scientists have known for years that if you stimulate specific parts of the brain, you will see an appropriate bodily alteration and sensation and can even induce illusions. This is in part how they have identified and mapped out the specific parts of the brain that are responsible for specific activities.

More importantly, inducing an alteration in body image in this woman had not addressed the "hard" problem of consciousness in relation to selfhood. It neither explained how her sense of self arises in the first place nor did it explain how millions of people come to describe their selfhood, consciousness, and awareness, *in short, themselves, not a duplicated virtual body image,* as fully separating and functioning in a state of lucid hyperconsciousness outside of their body.

Using this case of bodily distortions, Dr. Blanke had decided to discredit and dismiss the reports of millions of people, not to mention reputable doctors like Tom Aufderheide, Richard Mansfield, Douglas Chamberlain, and countless others, without ever even properly investigating them. Blanke hadn't even conducted a systematic study into what people with the experience of separation actually go through. Yet he claimed to have solved it. In reality, he had done nothing more than to simply attach a label, which reflected his own personal beliefs.

As we have read, those who enter the grey zone and experience separation from the body describe a strong sense of weightlessness and liberation from their body, with continued *external visual awareness* of events through a bird's-eye view—*in all directions simultaneously*—as if seeing in 360 degrees. They recall correct and objectively verifiable information, such as details of conversations or what their families or doctors had been doing and *even thinking*. There is greater *lucidity* and clarity of thought with reasoning, and expansion of consciousness, accompanied by a sense that their body has been shed—even though their selfhood still feels tethered to it by a type of "cord." *Importantly, their selfhood and consciousness are not felt to be the same as their body. Instead, they are felt as separate from the body.* In contrast, they don't describe their arms and legs changing shape, or seeing an image of their legs and trunk in their mind, or sinking into the bed, or falling from a height, or feeling that their legs may be coming to hit them in the face. *They also don't describe seeing a duplicate, a virtual image of themselves, which they recognize as unreal, acting independently in a distance, as people do during autoscopy.*

On the other hand, people with visual hallucinations and illusions, like the woman whose case Blanke had highlighted, or those with autoscopy, recognize that their real self is not the same as the distorted body images, visual illusions, or duplicated virtual image they may be viewing of themselves. *They recognize their experience is indeed a distortion and not real.*

The young man who had jumped out of the window and the two women whose autoscopic experiences the two groups of Italian scientists had studied *all realized that they were seeing an illusion, a clearly unreal and bothersome virtual body double. None described their double as their real self, even though they were confused and bothered by it.* They were in fact seeking medical help to rid themselves of the annoying and troubling duplicate.

A key differentiation between any hallucination and what people report experiencing in relation to death is that *people who experience hallucinations maintain an awareness either during the experience or after their hallucinatory experience has ended that it was not real, no matter how real it may have felt during the event.* Conversely, as we have seen, after the event, people with recalled experiences of death unanimously report that what they had experienced in death was "more real" than any real-life event, such as a wedding, starting a new job, a graduation, the birth of a child, and so on. This unwavering realization of having experienced something beyond "real" stays with them for as long as they live and distinguishes the experience of death from imagined experiences, such as ordinary dreams and hallucinations.

The recalled experience of death is unique. Nothing else comes close to it. To better appreciate how it is different from hallucinations, we also need to examine the testimonies of people living with hallucinations.

MOLLY BURKE IS A YOUNG, LOQUACIOUS, AND EXUBERANT TWENTY-nine-year-old Canadian woman who lives in California. She, like many other women of her generation, enjoys fashion, beauty, home décor, and adventure. In fact, she has a YouTube channel where she talks about these interests frequently. However, there is one thing that distinguishes Molly from most other women of her generation.

At just four years old, Molly was diagnosed with retinitis pigmentosa—a rare genetic eye disease that damages the light-sensitive layer in the back of the eye (the retina) and gradually makes people blind. Molly, now an adult, is blind. She lives in total darkness aside from perceiving shadows and light. She explains, "When I was eight years old, something very scary happened. I started experiencing visual hallucinations. [Suddenly] I began seeing all these intense, vivid, crazy colors. The colors were mismatched, and I could not tell what color was what and [they] were [constantly] flipping [around] each other. I was seeing stars, shapes, and geometric patterns. It was terrifying."

Molly was in third grade in school at the time. Although she could tell those images were not real, she was not used to them, and they scared her. She recalls a teacher who mocked her for drawing a tree with a multicolored trunk and leaves, just the way she experienced them. The teacher thought she was doing it for attention, but Molly was literally drawing what she saw. She explains, "It interrupts your vision and so you can't see properly. [That is because I] was seeing this visual hallucination and as a result . . . I couldn't see [the real] colors, and I couldn't figure out what color was what."

Molly has Charles Bonnet (pronounced bo-NAY) syndrome, a kind of visual hallucination that affects people who are going blind. It can happen after retinitis pigmentosa, but also after age-related macular degeneration, stroke, diabetes, glaucoma, cataracts, and optic neuritis—in essence, any condition that causes vision loss. Scientists think that the eye is no longer able to convert light into a visual message, and as a result the brain gets confused and creates images to make up for what it is missing. Sometimes, this can lead to simple hallucinations, but at other times they can be quite elaborate. These can include images of animals, buildings, even people. *Eventually, those who suffer with it get used to the visual hallucinations and carry on with their lives by ignoring them because they know they are not real.* Molly explained, "There are multiple types of hallucinations," and we have "zero control" over when they come on. For example, "I see blue, green and purple" colors, but for others it is different. Some people see elaborate "realistic full images of landscapes, trees, waterfalls, mountains, even historical figures dressed in historical outfits."

Throughout her life Molly has experienced vivid hallucinations of animals and people, sometimes even in historical costume, *but after experiencing those elaborate hallucinations, she knew they were not real.* This is the key difference between hallucinations and real experiences. Like those with autoscopy, she had become used to her hallucinations, and she could often distinguish what was real from what was not real, even as she was experiencing hallucinations.

Molly explained, "People [like me, who are living] with Charles Bonnet syndrome, know that our visual hallucinations are unreal. We are grounded in reality, and we stay present in reality the entire time [when] the visual hallucinations are happening." Even her doctors, finding that they couldn't help her, had said, "*Just try to enjoy the ride*," and that is what she tries to do.

IT IS CLEAR THAT WHAT SURVIVORS WITH A RECALLED EXPERIENCE of death report and their doctors tried to describe when using the term *out of body* is not what Dr. Blanke has labeled an "out of body experience." In fact, as we have seen, the only commonality between the two is that they both describe an experience related to the human body.

Perhaps Dr. Blanke and the people at the scientific journal *Nature* who reviewed his work had not come across testimonies of a recalled experience of death. More likely, though, Dr. Blanke, like others before him, had taken an opportunity to push forward a personal opinion, rather than making conclusions based on objective scientific studies.

This is a crucial point. Science needs to base itself on facts and data gathered through rigorous, impartial scientific enquiry. Otherwise, we are left with personal opinions and philosophy dressed up as science, using scientific-sounding language. Opinions and beliefs are important, but they are not science.

While we are all respectful of other people's opinions, every scientist knows it is wholly unscientific to relabel and "brand" one thing as something else based on the personal interpretation and views of one person, using one single case. This is why personal opinion and inherent preconceived biases, which shape people's interpretations of what they observe into what they want to believe, become so dangerous in science. Science is concerned with discovering the *true reality* of a given phenomenon, not what an individual or group of individuals—whether scientists or otherwise—may wish to believe.

In many ways, this was classic spin—not in politics, where people are accustomed to seeing it, but in science, where people don't expect to see it. This was a typical example of how you can dress something up as much as you want, but it still will not change what it really is. Or as the popular expression goes, "*You can put lipstick on a pig, but it is still a pig.*"

Dr. Blanke's effort would not be the last, or even the most incredulous, attempt at "putting lipstick" on a proverbial pig. In fact, after his marketing exercise, a handful of others followed suit and decided to arbitrarily label the experience of separation with death as a form of "autoscopy."

Originally, of course, the term *autoscopy* referred to a specific type of visual hallucination, *seeing a duplicated virtual image of oneself at a distance*, as the young man's case in Zurich and the two Italian cases illustrated. Progressively and arbitrarily, a handful of individuals labeled a wide array of other inexplicable human experiences as "autoscopy," too. This ranged from the feeling of a presence, like Shackleton had experienced in the Antarctic, to sensing a deceased loved one during bereavement, as Alderson-Day and Smailes had suggested. *Now some arbitrarily added the sense of separation from the body with death to what they wanted to label "autoscopy."* This was purely speculative and without evidence from rigorous scientific experiments. This again shows how scientists sometimes try to oversimplify complex human experiences into things they do understand by lumping them all together, even if that may not be correct. The reason for this is complicated and relates to the way the human mind works when confronted with new information that it can't process or explain easily (I will discuss this further in chapter 17).

In Anil Ananthaswamy's book, he dismissed "[the out-of-body experience as] an illusion [that is] a product of a brain that fails to correctly integrate all the signals from the body." This was despite the fact that *no scientist—not even Nobel Prize–winning ones—had shown how any experience, let alone one as complex as a vast sense of lucid hyperconsciousness with death, can arise as "a product of a brain."* He continued, "Despite their vividness, [they] *are hallucinations* caused by malfunctions in brain mechanisms." His justification? An anecdote.

Ananthaswamy described how Dr. Brugger had had him wear virtual reality goggles on a visit to University Hospital Zurich. The doctor walked behind him, filming Ananthaswamy from behind and projecting the image into the goggles. Instead of seeing the corridor ahead of him, Ananthaswamy saw himself, filmed from behind, walking down the corridor. He said, "The setup didn't quite work . . . but I did feel weird walking around watching myself from behind."

Ananthaswamy explained that Dr. Brugger had first attempted to carry out this experiment on himself in 1998. He had worn "goggles for an entire day, and had someone walk about twelve feet behind him, filming him with a video camera." Brugger recounted, "It was extremely strange. I lost the sense of where I actually was," before explaining that while wearing the goggles, "[I felt as though] *I was where I saw the action, rather than where I was actually executing the action*."

Based on this anecdote, Ananthaswamy concluded: "Brugger was having an out-of-body illusion." Why? Because, he says, "the sense of where he [Brugger] was located had shifted several feet, from being in his physical body to being in the virtual body." This, of course, was untrue. Brugger had not described his selfhood as being "shifted" in any way, nor had he experienced a "virtual body," nor a sense of separation from his body—i.e., being "out of body." *Throughout the experiment, he recognized that his selfhood remained where his body was situated, not where the camera was*. This was a simple camera trick, not an "out of body" experience, and was mislabeled as such in support of a personal belief.

Ananthaswamy clarified, "Brugger never actually performed the experiment in a rigorous laboratory setting and so never published the results, though it did get mentioned in an article in *Science*." He was referring to work by Dr. Henrik Ehrsson, a postdoctoral fellow at University College London in the mid-2000s, who placed goggles over volunteers' eyes, had them stare at a projection of their backs for several hours, then mimicked assaulting "them" (in reality, the camera) with a hammer. The volunteers in the study became startled because they were tricked into believing they were being attacked.

The experiment *didn't induce the sense of separation from the body that people recall with death*. Nothing even close to it. Yet when he published his results in 2007, Dr. Ehrsson proclaimed that he, too, had reproduced an "out-of-body experience" in the laboratory. Except he hadn't. He had also labeled a visual illusion as an "out-of-body experience," in accordance with his own beliefs. This was, again, without even considering conducting a systematic study into what people's experience of separation with death might be like.

The *New York Times* ran a story about Ehrsson's experiment, with a headline proclaiming "Scientists Induce Out-of-Body Sensation," and the media accepted the doctor's findings. The piece claimed the virtual reality experiment provided an explanation for phenomena usually ascribed to "other-worldly influences."

One can only assume that all these individuals genuinely believed what they were saying. This is why it is so dangerous when scientists and journalists—*people who are meant to be impartial*—do not consider the impact of their own inherent biases when tackling an issue, especially one as complex as consciousness and what happens with death. Yet the arbitrary labeling of the recalled experience of death as hallucinations by some people has continued and has become even more elaborate.

Chapter 16

TRUTH IS IN THE EYE OF THE BEHOLDER

In September 2023, the English magazine *The Spectator* covered the "renaissance" in psychedelic drug research over the last twenty years. "Not since the 1950s and early 1960s has there been so much interest in researching the therapeutic potential of psychedelics." The high point being the US Food and Drug Administration's approval of a "ketamine derivative for medicinal use in 2019" and giving "both MDMA and psilocybin (the psycho-active ingredient in magic mushrooms) 'breakthrough therapy' status, putting the drugs on a fast track to approval in the US, with the UK likely to follow suit."

Hallucinogens are drugs that can cause a person to see, hear, or feel things that aren't really there. Some are derived from natural substances, such as mushrooms (psilocybin, or "magic mushrooms"); ayahuasca, a tea native to South America (DMT); or other plants (e.g., cannabis). These are usually used in their natural form. Others, such as LSD, MDMA, and ketamine, are typically created synthetically in laboratories. The therapeutic potential of hallucinogenic drugs like psilocybin, which has been shown to have a dramatic effect in people suffering from major depressive disorder, has led to great hope and excitement among scientists.

Natural psychedelics such as psilocybin and DMT have been used for centuries by various cultures, especially in certain shamanistic practices. They induce altered states of consciousness. While some people can have very frightening experiences, others may report a profound sense of connectedness with the universe, a feeling of losing their self, together with seemingly "otherworldly" experiences.

Despite their ancient shamanistic roots, the recreational use of psychedelics in modern times is a relatively new phenomenon that really came to the fore in the 1960s (think of those images of hippies dancing at Woodstock). However, safety concerns with these drugs led to President Nixon's War on Drugs and their prohibition in the United States in the 1970s. Similar bans in other countries followed and terminated almost all medical research on these drugs for decades, until the early twenty-first century.

Amid the growing interest in the therapeutic effects of these drugs there has also been interest among a small group of scientists in their mind-altering properties, or "trips." In 2018, in an article for the BBC, Ben Bryant highlighted this with a bold headline: "A DMT Trip 'Feels Like Dying' and Scientists Now Agree."

Bryant recounted the experience of Iona, a participant in a study headed by Dr. Chris Timmerman and his colleagues at Imperial College in London. After having her vision blocked with an eye mask and being asked to listen to soothing background music, the hallucinogenic drug DMT, which some people call "spirit molecule," owing to claims that it can induce so-called "otherworldly experiences," was injected into Iona's veins. Bryant explained, "The hallucinations then hit her like a hurricane, as a sense of dread enveloped her."

Iona herself said, "My eyes were closed, but there was so much going on that it was really hard to focus. The one image I remember was lots of books opening and rainbows shooting out of them." She continued, "I felt this quiver. The only other time I've had it is when I was giving birth. [I had a feeling that] I'm not sure I want to do this, but a sense of no turning back, you're here and you've got to go through this. I don't remember my body being around after that. My body just didn't seem relevant anymore. I felt like I arrived in some '*consciousness soup*' which seemed like a different realm to the one I ordinarily inhabit—even in dreams. *It just seemed like everything was rotating and swirling and spiraling.* It didn't seem like there were normal space-time proportions going on." Iona and thirteen other volunteers then completed the "near-death" questionnaire, a test designed in 1983 by Dr. Greyson to identify near-death experiences in survivors of life-threatening situations.

The DMT subjects were asked to reply "yes" or "no" to a series of non-specific questions, such as "Was your experience unearthly?" or "Did you feel peaceful?" or "Did you see a light?" or "Did scenes from your past come back to you?" even though this test has never been validated or designed for use in these circumstances.

After analyzing the results, Dr. Timmerman reported that there had been "a strong fit on most of the elements" between the DMT responses and those with a so-called near-death experience. The "scores induced by DMT are comparable to those given by the 'actual' NDE [near-death experience] group." In his article, Bryant clarified, "The results were better than they had hoped."

Having read Iona's DMT hallucination account and the results of their study, many of us were baffled as to how researchers could have possibly found Iona's experience of a spiraling "consciousness soup," and others like it, to be the same as the dying experience. Perhaps because she had said, "My body just didn't seem relevant," they had assumed this was more or less the same as the sense of separation that people report in near death. *However, a feeling that one's body had not seemed relevant for a temporary period of time is not the same as the sense of separation that people report at death.* Science requires rigor, and their conclusions didn't make any sense.

Dr. Greyson's questionnaire had sixteen nonspecific questions, and responding yes to any four of them would lead to a false-positive result.* For instance, since hallucinogenic drugs lead to highly unusual experiences, most people would have to reply "yes" when asked if their experience had felt "unearthly." These drugs often lead people to see neon lights, lasers, and so on, so they would have to reply "yes" to any question about seeing a "light." Many also said "yes" to having experienced moments of peace and seeing a life event. Therefore, these volunteers were deemed to have had something resembling a "near death" experience.

Not only had Dr. Timmerman and his colleagues misused Dr. Greyson's test, but they had also concealed their findings regarding what people with drug-induced experiences really experienced by not providing a single testimony from any of those thirteen subjects about their experiences. If the DMT testimonies were so similar to the recalled experience of death, why were those testimonies hidden?

Thankfully, the BBC journalist had managed to talk to Iona. Her experience of "dread," seeing "books opening and rainbows shooting out of them," or a "spiraling and swirling consciousness soup" was drastically different from the recalled experience of death. Yet by concealing those people's testimonies of drug-induced hallucinations and disclosing only the "false positive" scores, Timmerman and his colleagues had managed to mislabel those hallucinations as the same as the dying experience. Of course, once more, many people, including journalists who read about these studies, had insufficient understanding of the subject, and so fell for it.

---

* The issue of false positives can be better appreciated through an analogy. Should a group of postmenopausal women on hormone replacement therapy (HRT) take pregnancy tests, some of those tests will come out positive even though they had not been pregnant. This is what is meant by a false positive result. Now, imagine if some scientists (who happen to vehemently believe that postmenopausal women can become naturally pregnant) boldly tell the media they have unearthed "scientific" evidence—*a pregnancy test*—that confirms their belief. When others point out their error, they reject any concerns, strongly disagree, and point to the fact that the *pregnancy tests had been positive as evidence.* Those in the media who don't know any better take what they say at face value and report on a "scientific explanation" and breakthrough "evidence" that shows how natural pregnancy is possible in postmenopausal women! Some scientists have been using the Greyson Near-Death Experience Scale from 1983 in situations for which it was not designed. This leads to false positives (e.g., after using hallucinogenic drugs). However, even though the contents of people's testimonies are obviously completely different, those scientists keep insisting that the experiences are the same.

Dr. Timmerman's claim was just another example of *putting lipstick on a pig.* Like Blanke, Ehrsson, and many others had done, these researchers used their personal opinions and false positives to claim people's recalled experiences of death are nothing more than imaginary experiences, like dreams or hallucinations.

To illustrate this further, in one study from 2019, a team of researchers claimed these experiences are nothing more than an imaginary disorder that arises in relation to falling asleep. The evidence for this claim? Interviews with individuals who said they sometimes felt a presence in their bedroom, saw spirits or demons at their door, or experienced sleep paralysis. These experiences do not resemble any aspect of a recalled experience of death, yet the researchers arbitrarily claimed these were like a recalled experience of death. They claimed these showed how the experience of death is nothing more than a sleep-related disorder, a trick of the brain.

A test with the power and accuracy to distinguish between people's testimonies in relation to death and other diverse human experiences, such as dreams or drug-induced experiences, requires much higher power of detection and specificity than Dr. Greyson's simple questionnaire, which was developed using only sixty-seven testimonies and was never tested in other circumstances. Instead, a test with such accuracy would need to be created using thousands of people's recalled experiences of death, dreams, drug experiences, and ordinary day-to-day experiences, using sophisticated and powerful mathematical tools such as artificial intelligence and machine learning, which were not accessible to Dr. Greyson when he created his test. *This is why we realized it was important to address this debate using an impartial method that would give mathematical certainty.* As explained before, we clarified with 98 percent certainty, using machine learning and many thousands of testimonies, that the recalled experience of death is different from drug-induced hallucinations and dreams.

Thankfully, with the resurgence of interest in hallucinogenic drugs, more and more people have provided testimonies about their hallucinogenic experiences—and scientists and journalists alike have started to document these in the scientific and lay literature. There is a striking difference between the way people describe drug-induced trips and how survivors describe a recalled experience of death. The drug users talk about swirling colors like tie-dye or being in an environment with fake grass and multicolor tables and chairs. They see disembodied faces or blades of grass that open up like concertina fans. They see dancing seductive figures or impossible creatures made up of multiple animals. In short: anything the brain can dream up, they see.

This illustrates the crucial difference between a drug-induced trip and a recalled experience of death. There's a reason we have been able to break it down into a consistent narrative arc that broadly involves separation, reassessment of life followed by a return, and fifty elements within this arc: because people experience some variation of the same unified and consistent themes. In drug-induced trips there are endless haphazard variations and differences between experiences. People returning from a drug-induced trip might report seeing moments of their life or family members, but *the context is totally different*—in exactly the same way that seeing a family member or some part of your life during any of your own dreams is clearly different from the testimonies we have studied together. In the grey zone, people don't just see random flashes of objects or places and people; they reevaluate their entire life from the perspective of morality and ethics. I have yet to read an account of a trip that included that experience.

These cases illustrate the challenge with language and the limitations of making inferences simply based on a few words taken out of context. People may experience seeing places, objects, and people, or being out of a room in multiple circumstances, including a dream. It doesn't mean that every time we think back or see people from our past, it is a dream. *By relying on words alone without context, it is exceedingly difficult, if not impossible, to claim any experience is the same as or different from another.*

Semantic precision is important in scientific study. Before completing our large-scale study using natural language processing, our team decided to dig deeper to see if we could find similarities between drug-induced experiences and the recalled experience of death. We examined every single scientific study—a total of nine had been published between 1994 and 2019 (this is when we carried out this search)—with testimonies of DMT or ketamine hallucinations. (We included ketamine because some people have claimed this drug also mimics the experience of death.) We wanted to better understand what those experiences were really like.

We thought we might find some similarities between DMT or ketamine hallucinations and the recalled experience of death. *However, we did not.* While the recalled experience of death narrative arc contains uniform elements of a separation from the body, evaluation and reappraisal of conduct in life, returning to a place that feels like home, and a decision to return to the body, as already mentioned, hallucinogenic experiences typically involved haphazard geometric shapes, spirals, and unrelated beings.

Overall, the reported DMT and ketamine themes could be categorized as: (1) *physical bodily changes* (like twitching or losing control); (2) *seeing aliens, spaceships, and other beings* (for example, elves; these beings are very different from the luminous, compassionate, loving being with enormous magnitude who guides people through an evaluation of their lives with humility); (3) *feeling of turning into animals*; (4) *seeing geometric shapes, waves, neon lights*; (5) *describing an alien dimension aside from the physical realm*; (6) *using "spiritual"-sounding key words*; and (7) *becoming egotistical.* There were also *many other miscellaneous, varied, and unrelated themes.**

To illustrate what I mean, I have included some examples of the main themes and their corresponding memories:

1. *Physical bodily changes.* One subject said, "[After handing the pipe over to a friend] I couldn't feel my body anymore . . . everything goes dark . . . I feel like I can't breathe. Whatever is left of my body isn't functioning. I curl up and force myself to take gulps of air."

---

* The full results of this analysis were published in May 2022 as part of the "Guidelines and Standards for the Study of Death and Recalled Experiences of Death" in the *Annals of the New York Academy of Sciences.* The full category of many more haphazard themes than could realistically be discussed here without becoming overwhelming (with examples) can be found in Table S3 in the supplementary materials.

2. *Seeing aliens, spaceships, and other beings.* Another person explained: "The [entities'] faces were the most disturbing aspect. All three had grey metallic faces. These were nothing like the Zeta-Reticluan Alien grays with large black almond eyes many people describe."
3. *Feeling of turning into animals.* One subject described: "My hands were complete monkey, along with big hairy arms, I felt like I was actually a monkey, and I thought I had begun to scream like a monkey, Ooh Ooh, Ee Ee, Ah Ah!!!!"
4. *Seeing geometric shapes, waves, neon lights.* Someone else said, "I was enveloped in an overwhelming sea of colorful geometry which resolved into a pathway leading to a very mechanical/alien-looking Mayan-themed pyramid." Another person shared, "I watched the kitchen units dancing before my eyes, sliding along left to right in outline form, neon colored reds, blues and greens."
5. *Describing a dimension aside from the physical realm.* Another person said, "I knew that the spiritual realm, the infinite realm, was and would always be the more beautiful and more intelligent realm." Although on the face of it this statement might sound like a recalled experience of death, when you examine the context of what is really being said it becomes clear that it is very different. This person continued: "*I knew this, because the aliens knew this.*" It also highlights how often people use words that might at first glance seem similar to what people say after their death experience, such as describing a different dimension or spiritual-sounding terms, but when these are put into context, it becomes clear they are talking about something different. In this instance, this person had really seen a world of aliens.
6. *Using "spiritual"-sounding key words.* This was seen in the previous example. Another subject described how "dancing colors and patterns passed before my eyes and [led] my gaze to the roof where I witnessed a perplexity of light and color twirling forever within each other, not unlike the murals and paintings of Angels that we see in Churches." Here again, this person had mentioned words

with a religious tone, but the context is clearly not like the recalled experience of death.

7. *Becoming egotistical.* This was often seen as exemplified in the following testimony: "The others started murmuring in agreement from behind me. 'You look Christ-like,' he said. I accepted it without question, it was obvious to me in my state. I was Christ. I was God." Here this person too was using religious terms, but the context was completely different. He had become quite egotistical and grandiose, which is not a part of the recalled experience of death.

This highlights one other key distinguishing feature between drug-induced experiences and the recalled experience of death (aside from the fact that the themes are different and that, after hallucinations, people realized what they had experienced had not been real). This other feature relates to the effects of the experiences on people. Whereas drug-induced hallucinations induce a sense of grandiosity (for example, people thinking they are God) during the recalled experience of death, there is no ego left. This leaves people to experience a profound sense of humility. Even the highly luminous, compassionate entity with great magnitude, power, knowledge, and wisdom is always characterized by complete humility and without any ego.

This sense of complete humility without any egotistic traits is reported in spite of the fact that during day-to-day experiences, none of us truly know what it is like to be without any ego, and either way we all have egos with greater or lesser intensity.

DREAMS, LIKE DRUG-INDUCED HALLUCINATIONS, ARE ANOTHER THEory put forward to diminish the significance of people's recalled experience of death to nothing more than an imaginary experience. After all, dreams are often full of fantastic visions, encounters with strange creatures, and adventures in lands and environments that seem strange and alien to us. Of course, at first glance and viewed from a neutral observer's perspective, it is understandable why, *despite what the people who have undergone this experience claim*, scientists like Dr. Borjigin and others may doubt the validity of their claims of being truly separated from the body, and instead consider their recollections to represent vivid, yet unreal dream-like experiences. Undoubtedly, anyone could have a fantastic, intense, and incredibly vivid imagined dream, in which they observe their own body from above and feel like they are floating and seeing things happening around themselves. Our AWARE-II and subsequent large-scale natural language study finally showed that, scientifically speaking, people's recalled experiences of death are not like dreams.

Certainly, when I first heard about these experiences and started to study them, I too considered these claims to possibly sound like an imaginary experience. That was largely because I, like everyone else, tried to make sense of them based on what I had previously known and understood. After all, critically ill people can become confused and may imagine all sorts of things. So why not this? I am therefore sympathetic to anyone who might think this way. However, with time and much greater study, I came to recognize that there were too many issues and unanswered questions that did not fit this assumption.

First, it would seem exceedingly odd and highly improbable for humans to somehow all have the same "dream," or imaginary experience, with the same particular focus of seeing their own body with death. Why should death trigger more or less the same experience in everyone, and why would, among all the possible themes that people could experience, they all somehow come to experience seeing their own body and all around it through a sense of knowing—in 360 degrees—so lucidly and clearly from outside, as well as reappraisal of our life viewed from the perspective of morality, followed by a return? We all dream, and our dreams are by their nature very different and diverse. This is one of the reasons why our research, even using objective mathematical tools with natural language and artificial intelligence, could so effectively distinguish between dreams and a recalled experience of death. Even when critically ill people have imaginary experiences, those experiences are somewhat different and diverse. So what is it about death that makes it different, and why do we all have more or less the same unified recalled experience, irrespective of culture, religion, and background?

Second, people universally describe a recalled experience of death as more real than real life. This hyperreal experience contrasts with dreams, hallucinations, and other imaginary experiences, all of which people accept as unreal once they awaken or recover from whatever was causing them to hallucinate. By contrast, people have a sense of consciousness and awareness with lucid thought process that is not confused and fading. This is what would be expected, since sophisticated brain monitoring systems also show that brain electrical waves slow down and then transition to a flatline state during a recalled experience of death. Yet somehow this flatlined dysfunctional brain milieu allows people to undergo a paradoxical vast expansion of their consciousness.

Third, individual elements of a recalled experience of death might seem dreamlike, but when we take the specifics together, this explanation doesn't work. For example, people might experience the sensation of floating in a dream, perhaps in the air or on a lake. However, as we have seen from the many different testimonies, the perception of floating in relation to a recalled experience of death is very specific. While "floating," the individual can also view real and verifiable ongoing events around their own body and beyond and simultaneously gain real and actual information about those events. In other words, this is not just a simple nonspecific feeling of floating. Instead, it is a very specific perception of seeing one's own body and events related to it, which are sometimes corroborated. Unless all the doctors' and nurses' testimonies have been untruthful, or they, too, had been imaginary, which is exceedingly improbable, if not impossible, then those people's testimonies cannot be considered dreams or imaginary experiences.

Fourth, building on this point, how have unbiased and neutral observers—doctors and nurses—been able to confirm the validity of what people claim to have seen and heard during these remarkable "dreams"? This makes them unlike ordinary dreams, or imaginary experiences, which by definition describe events that have not really happened, and so others cannot corroborate them. If someone is able to describe accurately and correctly what you and I or someone else had been doing and saying, we cannot classify their recollection as unreal or a hallucination.

Fifth, how is it that during these "dreams" people perceive things they had never imagined or experienced before during ordinary life? None of us have experienced 360-degree vision in our lives, so how can we imagine experiencing this in a dream? Our eyes cannot detect everything all around us all at once, and our brain cannot process such a thing. That is why we can't even try to imagine it.

Equally, how many of us know, or have experienced, what it is like to undergo the vast expansion of our consciousness with lucidity that people are describing? We can't imagine that either. Why, then, would we all suddenly "imagine" an expansion of our consciousness in and around death when we have never been able to perceive or imagine such a thing in life? It seems very odd, improbable, and inexplicable. Again, we have to ask: What is it about death that brings this out in humans?

Finally, the perceived separation of consciousness and selfhood from the body around death represents just one major element of the unique recalled experience of death. Another unanswered question is: If this is an "imaginary" dream, or hallucination, why would it then proceed to a detailed analysis of a person's entire life with a focus on morality and ethics? This is different from the slew of random experiences that characterize the testimonies of people who have taken hallucinogenic drugs. How can a detailed review of someone's life, including their actions throughout life, be considered imaginary and a hallucination? If that were so, then every time someone discussed reviewing their own life, would we tell them, "You are hallucinating"?

Dreams and drug-induced hallucinogenic experiences are indeed haphazard, random, and dissimilar. They do not follow the same narrative arc as a recalled experience of death. A word or image out of context, like "I saw a bright light" or "it was peaceful," might suggest a link. However, both dreams and drug-induced "trips" have ultimately haphazard themes. As we just saw, when a person who has taken a hallucinogenic drug describes a different realm, they are usually talking about a realm like those of space aliens. Equally, when they use spiritual-sounding words, it is not in the same context as those with the death experience.

This is why, in the end, as already explained, we resorted to the objective power of mathematics, using natural language processing and artificial intelligence. Ultimately, objective scientific methods are needed to discover the truth of any phenomenon; otherwise we are left with people's own subjective interpretations based on their own beliefs. We finally managed to put an end to this discussion with mathematical certainty. We know some people will still keep pushing their own opinions, but that is no different from any other situation in life.

Section Five

WHAT IT ALL MEANS FOR US

Chapter 17

PUTTING IT TOGETHER

IN THE FUTURE, SO MANY OF OUR CURRENT ASSUMPTIONS ABOUT death will be challenged, just as our old ideas about the binary concept of life and death based on a specific biological point is being challenged now. The field of understanding about what it means to live or die is ever expanding, whether we like it or not. Our choice is whether we acknowledge and integrate these new ideas into our social construct or we reject them. And the truth is that mainstream science's rejection of the recalled experience of death to date, and refusal to consider that the "grey zone" is a place full of experience, meaning, and wonder, is a choice. Likewise, the belief that the brain is the seat of consciousness is a philosophical construct, too: there is nothing in your brain that you can study that will tell you if your "self-hood" is in this area of the brain or that one.

Just like people once refused to accept that surgical infections come from dirty hands, or the Earth orbits the sun, some influential people in science, medicine, and research have had little interest in considering that death might be reversible, or that "dead" patients might be revived hours after their heart beats for the last time. I don't blame them: I can't imagine what challenges the idea of reversing death might bring to the health care industry, or even what effects it would have on all our laws and society in general. Even a question as mundane as who will pay to cover extra hours of resuscitation attempts using far more sophisticated technology, or days on a ventilator, is enough to sweep a matter as big as this one under the rug.

Yet scientists, administrators, and doctors have to be at peace with the demolition of accepted orthodoxies. We cannot allow ourselves to fall back on comfortable ways of thinking simply because emerging evidence points to uncomfortable new ideas.

These new dimensions of knowledge and ability are all around us, and as more hospitals get modern lifesaving equipment, they will find themselves in the position of reversing death in patients who would have been pronounced dead just a few years ago. On November 3, 2019, thirty-four-year-old Audrey Schoeman, an amateur mountaineer living in Barcelona, got lost in a snowstorm. She and her husband tried to shelter from the extreme cold and waited for rescue. As the hours passed, Audrey fell into delirium, "talking nonsense" and acting strangely. She lost consciousness as the extreme cold lowered her body temperature. At roughly 2:30 p.m. she died, with her lungs and heart going still. Her body temperature continued to drop, eventually reaching 18° Celsius (or 64° Fahrenheit) when she was found.

The rescuers airlifted the couple to Barcelona's Vall d'Hebron Hospital, where they arrived at around 5:45 p.m. The doctors confirmed that Audrey's heart wasn't beating, she wasn't breathing, and her kidneys and other organs had all stopped working. By this time, she had been without a heartbeat for some three hours and was more dead than when most people are being taken to mortuaries in hospitals. However, Audrey was fortunate. Instead of declaring her dead, the doctors placed her on extracorporeal membrane oxygenation (ECMO), a technique that uses a machine to do the work of the heart and lungs outside of the body. As discussed, this is the best of the best of what is currently available at select hospitals to resuscitate people. ECMO continuously pumps blood out of the body through large catheters in the neck and groin and warms it up and enriches it with oxygen before pumping it back into the body again. Eventually, at 9:46 p.m., her heart restarted. Audrey had been dead for six to seven hours when she was revived. It was only a series of lucky breaks—extreme cold, modern equipment (available only in very select hospitals), doctors willing to go the extra mile—that saved her.

Today Audrey lives a very normal life by all accounts, with apparently no memories of her time in the grey zone. Her case clarifies something crucial about the study of death and dying: the boundaries between the worlds of the living and the dead are in constant flux. Every new development moves the demarcation point on the map and expands the borders of the grey zone further into what was once simply death. What would doctors have thought of ECMO for resuscitation fifty years ago? And what developments that we can't even imagine will doctors and researchers fifty or a hundred years from now take for granted?

SCIENTIFIC EXPLORATION IS LIKE MINING FOR HIDDEN GEMS: YOU have to be willing to risk a lot to pursue an idea that others scoff at. Thomas Cullinan, a South African building contractor, decided to try his hand at prospecting. He purchased a piece of land that was miles from the established mines and started digging. Cullinan had faith in his land, despite the derision of other prospectors. And after just over two years of digging at different parts of this otherwise empty land, he discovered the Cullinan diamond; at 3,106 carats, it is the largest gem-quality diamond ever mined.

In science, too, those who seek treasure need to be willing to dig in places others reject in the hopes of finding a gem. Most of the leads that they pursue will not end anywhere. However, those efforts will eventually open doors to major discoveries. Today, while many scientists continue to "mine" in more or less the same areas, meaning fields of science that have already been discovered, there are some who take risks and are willing to look "outside the box" to unearth new discoveries in areas that have been undiscovered and untouched. Without their efforts and the risks they take—as exemplified by Dr. Sestan's story—we cannot make fundamental new discoveries in science.

You also must be willing to have your ideas, your motivations, and your fundamental belief systems mocked and dismissed: Sir Isaac Newton, the father of modern physics, spent much of his time exploring a host of unorthodox practices, of which alchemy was perhaps the most notable. In fact, according to some historians, Newton was first and foremost an alchemist, and much of what he studied was so unorthodox that it would be classified as "occult" today. He believed what he was pursuing was worthy of exploration, but most of it did not lead to much. However, the scientific gems that he discovered in physics came about as a result of his overall open-minded pursuits, not in isolation.

To many people, my hiding of images—first on ceiling tiles, then on shelves, and finally using a tablet computer, in areas where people were being actively resuscitated—seemed unorthodox, and I openly acknowledge this. However, without this, I would not have discovered all that I did later. I was prospecting for a potential gem while faced with a new and uncharted scientific field of inquiry. Now I can look back on my early attempts and recognize how rare a gem we were looking for. First, only 5 percent to 10 percent of people who undergo resuscitation attempts survive, and only 2 percent of them will recall the sense of separation from the body. This means that to capture fifty cases of people who could recall watching events from outside their body—the separation part of the recalled experience of death—which we had set out to objectively test using hidden images, we would have had to study 50,000 individuals. The cost of such a study would be astronomical.

However, more important is what we learned with time. People who undergo a recalled experience of death do so through what can be best described *as a field of consciousness*. This can be analogized to an electromagnetic field that can carry and at the same time capture information. This is what leads to a sense of knowing that people refer to as "seeing" or "hearing." In the grey zone of death, the concept of "seeing" or "hearing" is not literal in the same way that we may ordinarily look at something through our eyes or hear through our ears, meaning they do not use the apparatus of the sensory system. So, hiding an image in a room for someone to maybe see it is somewhat of a childish and naive idea. While I recognize that now, it was nonetheless necessary and valuable when I was starting out.

Now, almost thirty years after I started, we have shown the world how to use sophisticated brain monitoring equipment to peer inside and see what is happening to the brain second by second. Simultaneously, we have also shown how to test for lucid consciousness while people are in cardiac arrest and stepping into the grey zone. The more than 2,500 subjects who participated in AWARE-I and -II together, some of whose testimonies you have read throughout this book, represent studies carried out by experts in neurology and neuroscience, anesthesia, intensive care, and emergency medicine, in twenty-five leading hospitals and medical institutions. We even managed to use the most up-to-date artificial intelligence methods with our colleagues at the NYU Center for Data Science. These techniques were beyond our wildest imagination just a few years ago.

The result has been a fundamental shift in how medical science views death, and the beginning of a shift in how the general public does, too. AWARE-II and other studies discussed in this book suggest that death isn't the "end" we all thought, but rather the beginning of a new process, and that the dying person is merely wading in the ocean of death, not drowning in it irretrievably, at least not for many hours. As an article in the *MIT Review* put it aptly: "Just as birth certificates note the time we enter the world, death certificates mark the moment we exit it. This practice reflects traditional notions about life and death as binaries. We are here until, suddenly, like a light switched off, we are gone. But while this idea of death is pervasive, evidence is building that it is an outdated social construct, not really grounded in biology. Dying is in fact a process—one with no clear point demarcating the threshold across which someone cannot come back."

I hope that people will begin to understand that when the brain loses global function just before or after death, this is less "brain death" and more brain hibernation of sorts. The brain has hours yet when full function could be restored after being lost. In the meantime, through the process of disinhibition, the brain pours all of its resources into activities that will maximize its chances of staying alive—namely, getting the heart to beat again. It also activates abilities that existed merely as potential, yet dormant, states. For instance, the genes that repair any damage to fetuses but are "turned off" at birth. In death, these genes flip back on, presumably to join the brain's battle to stay alive.

In the same way, as already discussed, when people enter the ocean of death, there seems to be an inflection point of brain dysfunction, which triggers disinhibition and activates certain functions that were lying dormant in a sort of "sleep mode." This provides access to extreme, yet otherwise hidden, capabilities in the depths of human consciousness that in turn give access to other realities that are now more relevant in preparation for this new state of being.

While the doctors and nurses fight to save the individual, the dying person's sense of their own consciousness becomes enormously vast: like the cosmos compared with the Earth. In this state of hyperexpanded and hyperlucid consciousness, people are filled with a deep and profound understanding of themselves and of life: they are liberated from their body yet have a hyperconscious awareness of all events around and beyond themselves all at once and in 360 degrees. They realize that their real self is their consciousness, not the body. In this new, expanded state, their consciousness and selfhood feels like a field of energy, analogous to an electromagnetic field, one that can penetrate the thoughts of others and objects. Yet people still feel connected to the body through a metaphorical cord of sorts.

Linear time loses meaning. Instead, people experience millions of realities, almost downloading them like computer data, simultaneously. They review and judge their life based on the quality of actions and intentions. They realize that there has been a cause for everything in their lives. They recognize that they are responsible for their own actions and intentions, and they relive the downstream consequences, or domino effect, of their actions on other living beings. They relive their own actions through the eyes of the other living entity, human or animal, and deeply feel how they felt in that moment. Thus, they appreciate the positive and negative value of their actions. *They also recognize that the value of their actions was determined by the intentions behind them.*

They realize that even seemingly small moments of genuine kindness, sacrifice, and selflessness are immensely valuable, even if they were not valued by others. The status and values of society are irrelevant. Instead, they judge themselves based on a universal moral and ethical code, independent of their belief systems. It is also unaffected by any self-justifications, such as how they may have blamed others in life to vindicate their own incorrect actions.

Now in this hyperalert and lucid state, where all actions and intentions are laid bare for themselves and others, whatever they have done is experienced far more intensely and profoundly. There is no room for dishonesty, and there is no disagreement with the judgment, as it is all carried out and experienced by the person themselves. The effects of any positive actions lead to a tremendous sense of euphoria and happiness while at the same time the pain of any negative actions leads to a tremendous sense of shame, embarrassment, humiliation, and remorse that culminates in indescribable anguish.

It should be highlighted that these sensations, such as humiliation, are experienced multiple orders of magnitude more intensely than how they may be experienced during day-to-day life. So, both the intensity of the euphoria is immensely higher but so, too, is the intensity of the pain of embarrassment, shame, and humiliation.

People describe going back to a place that feels like home, or where they belonged. Unlike their day-to-day experience in life, this place, described as "perfection," is infused with an indescribable atmosphere of goodness, kindness, and understanding. People recognize that their day-to-day life had been a first step in a larger overall process of gaining knowledge of truths that lead them to wisdom and that this process continues in this place. They also recognize they have been there before and have also lived previous human lives as part of an overall ascending development process.

There is a vast hierarchy of knowledge in this domain, far beyond themselves. They encounter beings in the form of light or a transparent human form, some of whom have a guiding role and assist the person to gain understanding. Those guiding beings have a much higher magnitude of knowledge and power than they (or others) do and enjoy a much higher position on the hierarchy of knowledge. Yet there is no ego to their power and knowledge. Instead, they are infused with humility. We found it astonishing that during the recalled experience of death common emotions that we ordinarily share with animals, including primates—not just ego, but also fear, dread, jealousy, anger, sensual and physical attractions, and so much more—were not at all seen, even though they are a part of our ordinary lives. In their place we found only positive attributes, such as kindness, generosity, benevolence, compassion, understanding, learning, knowledge, assistance, dignity, and, of course, humility. This further suggested that the recalled experience of death could not have been constructed in people's minds. By contrast, those primate and animal emotions were commonly experienced during imaginary experiences created in the mind, such as drug-induced hallucinations and dreams. With the recalled experience of death, there is also recognition that beyond the luminous, seemingly "perfect" guiding entities, a far higher originating source—with an unfathomably higher magnitude of knowledge and goodness—exists and is that from which everything originates. However, people cannot envision what this source is, as it remains far beyond their abilities of comprehension. Nonetheless, it is recognized as having a far higher magnitude than even the most powerful and "perfect" luminous guiding beings they encountered.

Ultimately, people recognize that their life had reflected a dynamic and ascending educational process, with the goal of assimilating universal ethical principles that can lead them to become a true human being and gain knowledge of truths. They recognize that in this regard, they had not been as good as they needed to be and had not achieved their ultimate goal.

They realize they need to become a better and more moral and ethical human being: this includes acting in a manner that is consistent with a dignified life. Even "good" people need to continue to improve in order to climb up the hierarchy of knowledge and understanding. This is how their actions and intentions in each scenario of life had provided opportunities for them to advance along this track. They recognize suicide is not an option as you cannot really die. If anything, it compounds the problems you may have been trying to escape from and creates far greater pain and suffering.

Everything in this new state feels far more alive and real than anything experienced previously in ordinary life. In fact, compared with what they are experiencing now, "real life" feels immensely less real. They feel detached from that life, which now feels like a dream by comparison. By and large, people do not wish to return. However, they learn that there is a purpose to their life, and they are told to return. They are given a second chance to accomplish what they were supposed to. This explains why after this experience, people are typically positively transformed and live with less self-centeredness, more altruism, and less fear of death.

Here's another twist of the recalled experience of death: in all our interviews and all our research, there is as much that we *never* hear as there is of what we do hear, which also suggests these experiences are not constructed in people's minds. We learn as much from what is not said to have been experienced as we do from what is recalled. We have never heard a survivor discuss experiencing the recognition that the sociocultural aspects of death—the religious traditions, language, or rituals—were found to be valuable. In all these testimonies that we have studied, many were from people who followed a religion. Yet none came back to say that "in my review I learned how important it had been that I was following all the rituals of my religion." *Everything simply boiled down to what they had done and the intentions behind them.*

If this experience was being constructed in the minds of religious people, then how come none of these other ritualistic or sociocultural aspects of religion, which they had so valued in life, are reviewed and analyzed in death? No matter how passionate a person was about their beliefs during life, in death the rules are different. They realize that the idea that "belief" in a religious entity alone is sufficient is wrong. What actually matters at the core of belief is how it directs your actions. Does it lead to selfless humanity, and the ability to put others ahead of yourself—be it your child, your friends, your patients, or simply strangers on the street—and does it drive you to gain greater knowledge? *In short, what comes out of this is that belief is important only in so much as what it led you to do in your life.*

It is also equally important to highlight that after they return, people do interpret and filter what they had experienced through their own personal belief systems, referring to the luminous, loving, guiding entity as Jesus, God, Mohammed, Krishna, angels, and so on. But as we studied so many diverse experiences, we came to realize these people were all talking about the same thing. A guiding being that is luminous, loving, and with enormous magnitude. The labels they used were attached afterward, but the essence of what was being experienced was the same, irrespective of what they called it.

Another facet of seeing the recalled experience of death through a religious lens is the realization that people didn't feel like they got away with their misdeeds. Nobody found that their misdeeds toward others were wiped off—forgiven away—simply because they believed in a particular religion or religious figure. Also, instead of a religious version of hell, survivors experienced a state of "hell" of humiliation and the pain they caused others when they relived an amplification of their own experiences and choices in life. If this experience was constructed in people's minds, they should have seen what they had been told to expect. In fact, as we discussed, some people have mistakenly claimed to have had hellish or distressing so-called near-death experiences and have referred to being attacked by demons and similar creatures. However, as we showed in the AWARE-II study, those terrifying memories are not the same as an authentic near-death experience or recalled experience of death. Instead, they are memories formed later, when people were waking up from their coma in the hospital. Because they were neither fully unconscious nor fully awake, they misattributed what doctors and nurses were doing to help them as frightening attacks by demons. I already mentioned that in the medical literature these cases are well described and called intensive care unit (ICU) delusions. Labeling them as "hellish near-death experiences" is not scientifically correct.

In life and in society, people value status. Yet in the life review, they do not dwell on their status in life. Why don't people who value money and status recognize them as being valuable in and of themselves during their experience of death? This and all the other features we have pointed out further suggest the recalled experience of death is not constructed in people's minds.

I BELIEVE THAT OUR NEW DISCOVERIES AROUND DEATH, INCLUDING the recalled experience of death and all the ideas and discoveries associated with it, can change the world. It's not easy to put your career and reputation on the line with ideas that so many reject. Yet these discoveries are only the beginning. The ripple effects from the recalled experience of death, and how it will change how we live, what matters to us, our priorities, and our philosophies, will be profound. In effect, what is experienced seems to be that next dimension, in a sort of four-dimensional or five-dimensional or multidimensional world: one in which what people experience through the grey zone is much more hopeful.

Here's another important point: I could not have done the work I've done if my mind was closed to the inexplicable or I was unconsciously looking to reinforce my existing belief systems. Likewise, Dr. Sestan could not have made his discoveries if he was not willing to actively evolve his thinking, as the data and the results of his study surprised him. In order to find those hidden gems, we have to set aside the old belief systems and mental shortcuts that we have taken for granted and relied on over the course of our lives and our careers.

Throughout this book, we have discussed much about how people have often dismissed the recalled experience of death. Now I want to explain more explicitly why this happens, as it reflects human nature and the limitations of our brain and mind in processing new information. The main reason we are so quick to dismiss or disregard the inexplicable can be found in psychology, and specifically our brain's preference for mental shortcuts, also known as heuristics, and reassurance that our own beliefs are accurate and true. Cognitive biases are a systematic deviation from rational judgment and are a feature of the way the human mind processes new information; they allow us to process a lot of complex information very quickly while also not becoming cripplingly overloaded. In order to do so, our mind works through inherent mental shortcuts. While these heuristics allow humans to quickly process new sets of information and make decisions relatively quickly without getting stuck in the details, they lead to systematic deviations from rational thinking and understanding of reality. In short, our minds do not process information in the way that a neutral computer would do. Instead, we act more like a computer that analyzes information through distorted software and consequently produces distorted conclusions.

Confirmation bias, the tendency to search for, interpret, favor, and recall information in a way that confirms or supports one's prior beliefs or values, is one such heuristic. And it leads us to routinely reject new ideas and information that don't fit with our own prior mindset. Instead, we mostly focus on what we already accept and believe and reject evidence to the contrary. If our hypothetical computer accepted and analyzed only new information that confirms the information it has already attained, and rejects everything else, it would clearly lead to a very distorted set of conclusions and projections. For humans, heuristics, or mental shortcuts, while necessary to get through our everyday activities, like choosing which car to buy or where to go on vacation, are clearly detrimental when it comes to understanding the truth that underlies complex, hidden layers of reality.

Another major type of heuristic is cognitive dissonance. This is the mental anguish that arises in our minds when we come across new information that contradicts our own actions, beliefs, values, and knowledge. To relieve our mental anguish—the dissonance—we have two options. We can try to take the time needed to reevaluate our own understanding of that subject by getting to grips with the depths of that new piece of information. That is hard and requires a lot of effort. The alternative and easier way to relieve the dissonance is to simply reject the new information as not being correct. When we reject something that doesn't fit with our world view and beliefs, we no longer feel the mental pain. It makes sense for us to reject things, as it again helps us process information quickly.

As a result of these heuristics, people systematically and routinely reject new ideas and information that don't fit with their own prior mindset. Instead, they mostly focus on what they already accept and believe and reject evidence to the contrary. If we pay attention to how we respond to new information in our lives, we should easily detect both confirmation bias and cognitive dissonance in ourselves consistently. These heuristics can lead us to comfortably accept many social constructs, including: "Everyone in my professional field says death is a binary? Well fine, I guess that's what I believe, too."

The greatest tragedy in all of this is how many people will die and remain dead who could be saved, just because it will take us years to shake off these constructs and begin to write new ones. Frequently, I see news stories about athletes who have been declared dead after a freak accident on the field. I have to wonder what their fate might have been like if people had truly applied all the latest discoveries to try to bring them back, or of course if their deaths had occurred just a few years in the future.

I do not blame people for being skeptical, at least initially. As mentioned, when I first heard about what I came to call "recalled experience of death" and started to study it, I, too, thought it sounded like an imaginary experience, perhaps drug-induced or resulting from some other prosaic cause. After all, many of us have had the experience of sitting at the bedside of an ailing loved one and watching them wave to or talk with people who aren't there or tell us things that clearly aren't true or even plausible. So surely the recalled experience of death is no more mysterious than that? I have some sympathy for anyone who might think this way. However, as I mentioned previously, with time and greater study I came to recognize there were too many issues that did not fit with this assumption and too many unanswered questions about the very nature of it.

Some of those questions I have discussed already as well as those core components of the recalled experience of death. Even the most convinced skeptic is tripped up by the simple universality of the recalled experience of death. We have interviewed people who entered the grey zone on a rural road in Iran, a busy street in London, a market town in Mexico, or a farm in the American Midwest. They have had these experiences at four years old, twenty-two years old, forty-eight years old, and seventy-five years old. Some died years before social media and the internet. Some are young people who live "terminally online" today. Some are devout, some agnostic, some atheist. No matter what their background, these vastly disparate individuals describe the same experiences, in similar terms, and using almost identical vocabulary. If a recalled experience of death is simply a "long, strange trip" or a dream, then how is it that a young student in the Middle East in the 1970s might dream the same dream as a world-renowned surgeon in America ten years ago? Or a child comes to share the same experience as a grandmother?

ALL THE STUDIES AND DISCOVERIES THAT WE HAVE DISCUSSED over the course of this book have the possibility of saving untold lives. First and foremost, Sestan's work has demonstrated that a person can be resuscitated hours after we previously thought their body would be too damaged to be saved. As this understanding spreads, it will encourage doctors and emergency personnel to continue lifesaving efforts, potentially saving millions of people who would otherwise stay dead. Audrey Schoeman owes her life to doctors who were willing to consider that her death was reversible. People who die of untreatable conditions will remain dead; however, there are also many people who die every year from reversible, meaning potentially treatable, causes. Think of all the young athletes who die on the playing field, or young healthy soldiers who die on the battlefield from exsanguination, or the civilians who die in wars, or others who are shot in other circumstances, but whose injuries could have been fixed with timely interventions if their hearts hadn't stopped. Yet once those people's hearts stop and they receive some period of CPR, they are all declared dead based on the incorrect assumption or belief that nothing else is possible.

Similarly, there are all the people who are declared dead from other injuries, such as after bleeding to death from a shark or other animal bite or suffering other accidents. Then there are all the people like Audrey who die in the cold but who were otherwise healthy, or those who drown in ice-cold water after their ship sinks and whose bodies have been fully preserved well into the postmortem period (think of the people who drowned after the *Titanic* sank, or all the other ships that were sunk and will sink in the future). There are also all the otherwise healthy middle-aged people who suddenly die after a devastating heart attack.

I think that eventually these individuals will be widely considered salvageable. But we first have to change how we define and understand death, and then accept that more is possible. Once we change our thinking about death, we can begin to do more for people after they die. Expanding the technology that Dr. Sestan and his team developed to make it a part of standard medical practice all across the world is the first step. But there is so much more that can be done to expand on his work, which is itself still relatively preliminary.

Another medical benefit of work in this new field relates to the ability to optimize organs in people for transplantation, as Dr. Robert Montgomery had been pursuing. We saw how he is now alive because he was able to receive a transplanted heart and how his own ethical dilemma led him to take an infected organ and find ways to treat it. As a consequence of this sacrifice and being alive, he has been able to give life back to so many more people through the innovations that he has developed in the field of transplantation.

One of the most incredible areas of work that he is now actively engaged with is xenotransplantation: the transplantation of organs from animals into humans. To do this successfully, scientists like him are carrying out ever more innovative studies by transplanting pig hearts and kidneys into people who earlier gave consent and who have now been legally declared brain dead and are kept on a life-support machine. In this manner, these scientists can see how the human body reacts to an animal organ before fine-tuning the science that will provide such organs for the masses of people across the world who are waiting for an organ and who will eventually die due to the scarcity of organs. This science has only been possible because of our new understanding of life and death.

I should also mention that scientists who paved the way for a new field of research by taking pieces of dead people's brains hours after death some twenty-five years ago are now seeing the fruits of their labor. Today, healthy, functioning minibrains the size of a fetus brain are grown from pieces of "cadaveric brains" in the laboratory. Scientists have even started to fuse these laboratory-grown human brains (organoids) with the brains of living animals to create a hybrid human-animal. When I recently discussed this with Dr. Alysson Moutri, a neuroscientist and stem cell researcher from the University of California in San Diego (UCSD) who is considered a world authority on the subject, he explained that incredibly those animals had not changed or developed human qualities. This was despite quite substantial amounts of human brain being fused with their own brains. This adds further intriguing questions to the already complex issue of whether the brain produces consciousness or simply mediates it like a television set. If an animal now has a brain that's, say, 50 percent human, then is it still an animal or is it a human? And if all the unique features of human consciousness are produced by the brain, as so many scientists think—rather than being mediated by the brain, as Sir John Eccles thought—then why didn't those animals become at least partially human? This new frontier will also change how we understand the nature of consciousness and what happens to us all when we die. To understand this new information of death, we have to be open to ideas that would have seemed absurd even a few years ago.

One element of the AWARE study is how the individuals in the grey zone experience the world around them. This lucid hyperconsciousness is not experienced the way we see and hear in our everyday lives. Instead, as we have already said, survivors with a recalled experience of death describe what can best be called a field of consciousness, perhaps similar to an electromagnetic field that can carry and capture information, and they feel the brain mediates their consciousness without producing it. As alluded to, this explains why I now realize hiding images in a hospital room for the dying individual to see was ultimately unlikely to be sufficient to explain the phenomenon. At the same time, they recognize that their consciousness, as they have understood it to be during their life, is merely the tip of the iceberg. Like the iceberg, there are huge and unexplored areas of their being; their experiences as a living person merely brush the edge of what it means to exist. That is why I have constantly referred to the concept of *a universe of consciousness* to contrast with the oversimplified idea that considers consciousness as a simple emergent property, like heat. These new understandings regarding consciousness need to be recognized as valuable data and incorporated into our science. No rational scientist should dismiss and throw out human data related to their field. How can we ignore these vast sets of data and still say we are studying the nature of human consciousness in an impartial manner?

WITH EVERY SCIENTIFIC BREAKTHROUGH THAT BUILDS ON THE last, it becomes increasingly difficult not to walk through the doors of knowledge and discoveries that are being opened up. If these scientific breakthroughs continue to push the bounds of death, then we need to question: When is a person dead or no longer able to be saved? What are life and death even? Is there a clear distinction between these two anymore based on some arbitrary biological time point? Should we not bring consciousness into considerations of life and death? Surely what we are all concerned with in terms of life is conscious life—whether that be at the beginning (in the womb) or at the end of life—not an arbitrarily chosen biological time point.

The moment Dr. Sestan's team placed the first pig brain into the machine marked the beginning of an experiment that would jolt the door open to a new era in science and unravel a confounding set of unprecedented questions for bioethicists, scientists, judicial experts, legislatures, and the public alike. Doctors have already inquired whether BrainEx's brain-preserving technology has a place in medicine. We can envision a scenario in which disembodied human brains could become guinea pigs for testing cancer drugs and Alzheimer's therapies deemed too dangerous to try on the living.

Would someone who has died and then been placed on BrainEx or OrganEx become conscious again and retain memories, an identity, or legal rights? Could researchers ethically dissect or dispose of such a brain? Can we naively assume that deceased people placed on OrganEx will not be alive and that we can just revive and then remove their organs for transplantation? Or will OrganEx actually cause society to lose its supply of organs for donation, since otherwise healthy people who die could be revived? What will happen to all those people who are dependent on receiving an organ transplant for their own lives? And what will happen when human brains grown in the laboratory from dead people are fused into the brains of living people suffering with brain diseases such as dementia and stroke? While the brain disease might thankfully be cured, what will happen to the recipient's consciousness and selfhood? Will they turn into a hybrid of the person who died and themselves, or will they remain as the same person they have always been? If consciousness is produced by the brain, then we would expect the person's selfhood to change. But if it is not, then we would expect the person to maintain their prior selfhood. Now it becomes even clearer why mainstream science can no longer distance itself from the impartial study of consciousness in relation to death.

Ethicists, including Drs. Nita Farahany and Hank Greely, who were at the NIH meeting in 2018, have continued to ask these profound questions—not just behind the closed doors of the NIH, but by pulling the curtain back for others to also weigh in. In an editorial titled "Part-Revived Pig Brains Raise Slew of Ethical Quandaries," published in the same edition of the journal *Nature* as Sestan's study, they wrote: "If researchers could create brain tissue in the laboratory that might appear to have conscious experiences or subjective phenomenal states, would that tissue deserve any of the protections routinely given to human or animal research subjects?"

These are questions for philosophers as much as scientists, and like the cave dwellers in Plato's famous parable, those who see a new truth for the first time are likely to be berated—or worse—for it. The cave dwellers who prefer to keep looking at the wall may deny that maybe someone broke free and experienced something they haven't seen yet. I and an ever-growing band of medical professionals, doctors, scientists, and ethicists are choosing to look away from the wall and toward the figures "in the light." We hope you will, too.

Chapter 18

THE FLIGHT OF LIFE

So far, we have mainly addressed the scientific side of our research. But an equally important aspect to this is its human dimension. This is something that we have only touched upon so far, but I think is important to address, as it will have the greatest day-to-day impact for all of us as human beings.

The loudest and most powerful message that shines through from the testimonies of the people with a recalled experience of death is one of hope, together with the fact that at the same time the truest gauge of who we are as human beings is the extent to which we have been able to develop our human dimension. Based on these testimonies, there is also recognition of a precise accountability and scrutiny with regards to every action and intention, gauged against what is a seemingly universal moral and ethical standard. In this context, what is the value system by which we live and conduct ourselves? How we live determines what our real worth is in the end. None of us possess anything else that is lasting, meaning that no one can take away from us. Beyond all else, what remains is who we are in terms of our true humanity and what we have accomplished in our time in this world in that regard.

While the message from the recalled experience of death is clear, the level of precision with respect to the review of life and what is required to be accomplished can be daunting—not to mention the cause-and-effect relationships and a recognition of the existence of a hierarchy of knowledge that reflects a developed field of perception capable of understanding truths and states of "perfection." What is being repeatedly understood is clearly something more than just what it means to be a "good person." We are seeing suggestions of what is in effect a curriculum of humanity—traits that need to be developed. And we see the requirement to achieve this is to be cognizant and attentive at all times with all our choices in life. Since in death we evaluate every moment of life, then it can be deduced that there is clearly no "off" time: every moment in life is a learning opportunity and there is no separation from that. It is active and ongoing while we are at school or work, and it is equally active when we are at home or in a social situation. *It is constant.*

The standard by which one is gauged is high. Yet at the same time, as many people's testimonies have highlighted, we are all human and make mistakes. As one of the survivors, Steve, highlighted, we can deceive ourselves and justify our incorrect actions, but in death we recognize their true reality. Equally, sometimes others don't treat us correctly. We saw how young Rachel Finch would go into fits of rage against her closest family. Surely, they must have done something to provoke her. Yet in her recalled experience of death, she highlighted her own errors, not those of other people, and this changed her relationship with her father. We may understand the cause-and-effect relationships that led someone to do something wrong to us, as Mary Neal recounted, but we are still accountable for our own errors and misdeeds in those interactions.

The cases and testimonies we have analyzed also highlight the practical and dynamic challenge of being a true human in the way that is evaluated at the end of life. There is no escaping from it. This I find more and more intriguing.

If we rise up to the challenge, we still have to acknowledge a lack of clear and definitive models that are suited to the times for how to practice those universal (as opposed to imagined or personal) ethics and morals to help develop our humanity. Different cultures and different people have divergent views about morals and ethics. So, what are the authentic standards to follow, and how do we apply them in our complex day-to-day life challenges?

We are not the first people to face hard existential questions. Humans have sought to understand the purpose of life and the nature of the self (whether we call it psyche or soul, as the ancients did, or consciousness, as modern scientists do) as well as what happens to our consciousness in death. Likewise, most of us strive toward morality and ethics. Our motivation to find these answers waxes and wanes with the vagaries of life; confronted with moments of loss, perhaps of a loved one, or when faced with a terminal illness, the need to find answers to these questions is reignited.

In the past, religions attempted to fill the deep and universal need for meaning in people's lives; however, their power has waned with the rise of the scientific method and rational thought. Despite this, interest in spirituality and transcendence—meaning the search for a higher purpose—has not waned, and in fact, continues to grow rapidly. This suggests that at the core of their lives people are still seeking a higher purpose to life, which neither modern society nor science has been able to address. We all seek these answers in different places, and from different sources, or as some people do, we just give up looking or show no interest in the first place.

No matter where we grew up, and no matter what ethical code we live by, we humans are all cut from the same cloth and share the same constitution. Likewise, everyone—whether rational scientists or otherwise—is affected by the same issues that shape our modern life. Even the most rational person will still seek answers to their deeper purpose and what happens to them with death. After all, when faced with that inevitable reality, and given a choice between continued existence or annihilation, few would voluntarily choose the latter. What would be the point of experiencing a flash of existence, only to be annihilated and disappear forever afterward? I confess that I am no different from others in that regard. Decades of being a scientist or having scientific training doesn't change that reality for any of us.

This conundrum—how rich and full our lives can feel, but how brief and insignificant they actually are when considering our life span of about eighty years compared to the vastness of millions of years of geologic time—affects us all. And most people, scientists or not, when faced with death would find it more meaningful to know that annihilation is not certain. The fact that even those who reject this notion are intrigued by it highlights its universal importance. That is why I chose to study this issue to the best of my ability, using the tool I thought I could rely on the most: science.

TOMORROW IS PROMISED TO NO ONE

I have always searched for a deeper, more meaningful purpose than simple success in my life. This has become more and more important over time, especially starting with my years as a medical student, when I first saw people departing their lives, like passengers on a journey that has come to an end. Seeing this has increasingly reminded me that my own voyage will inevitably dock for the last time. And as I depart from this journey, will I be wondering what it all meant?

This hit home for me during the challenging times when I—and thousands of other doctors and nurses—lived and experienced COVID-19 in New York. I was one of the intensive care physicians working on the front lines of the storm as it hit in March 2020. Very rapidly, we found ourselves short on time and running out of space. We were opening a new thirty-bed intensive care unit almost daily to try to deal with the influx of critically ill people. When all was said and done, we had a whole COVID-ICU building with almost three hundred ventilator beds because there were just so many patients flooding in on an hourly basis. Though we had been warned—by China, by Italy, by Spain—the whole city and state underestimated COVID's impact. We had known the storm was heading our way, but too many had simply hoped it would not come.

We managed an unprecedented number of "crashing patients," people in severe distress, panting frenetically, breathing at a rate of sixty breaths a minute, or one breath every second. They were exhausted, shattered, and petrified, with eyes bulging through fear. They were unable to breathe any longer on their own. However, due to the sheer volume of cases, people were unfortunately forced to wait for available ventilators. Given the number of crashing patients and limited resources at our disposal, we did our best to save as many people as possible. We were sending emergency response teams to put out fires in every corner of the hospital. The emergency pager and overhead calls for help were going off incessantly.

I can only compare the experience to a war zone. And like in a war zone, we were making life-and-death choices throughout our thirteen- to fourteen-hour days, for days on end, with few to no breaks.

It was disheartening to see rows and rows of people who looked like zombies hooked up to machines. I had to remind myself that they were indeed people—someone's child, father, mother, brother, sister, wife, husband, or friend. Though in reality they looked like husks of people. One out of five of the five thousand people who came to us in that first wave succumbed to the devastating complications of their disease.

Nurses and all medical ranks were asked to join what essentially became a quasi–national service. It was a conscription, plain and simple, just like the draft during a time of war. Staff members were notified where they would be needed, and they were expected to show up. Just like with an Army draft, you had some people who found reasons not to show up—because they were, understandably, scared to die. On the other hand, we also had people who voluntarily risked their lives to serve in any way they could. One of them, a world-renowned professor of ophthalmology in his sixties, showed up one day and after introducing himself said: "My wife thinks I am volunteering at a call center. She doesn't know that I have volunteered to be on the frontline of the ICU. She will kill me if she finds out."

That was because many doctors and nurses had already died on the frontlines. Every time we treated a COVID-19 patient, we recognized that person could cause us to end up on a ventilator and die within weeks or months, like so many others. On more than one occasion, to preserve someone else's life, I and others had to expose ourselves to forceful gushes of COVID-19–infected air being blown directly into our faces, sometimes for up to an hour at a time, with just a mask as our sole line of defense.

Imagine seeing all the complications—limbs turning into gangrene, blood clots, dialysis, sudden death—and then having to expose yourself to their cause. We didn't know how many of us might be alive in two months, but we certainly expected some not to be.

These events caused my colleagues and I to look at life more existentially. We were asking ourselves deeper questions about our purpose, knowing that our time might come to an end soon.

One explained, "[This] allowed me to ask myself, 'What is it that I genuinely needed to do? What is life about really?' Death is a completely normal process of life. The question is, what do you do in between?" Ah yes, the in between birth and death. This thing we call living.

The global pandemic touched nearly every corner of the world. It inflicted untold suffering on countless families and upended the very fabric of daily life, from how we interact and work, to how we raise and educate our children. At the same time, it brought to the fore larger and more difficult questions: What is the ultimate meaning and purpose of our lives? Is there some deeper meaning to our existence that transcends our day-to-day life? And what will become of us, and our loved ones, when we die? In that sense, the pandemic served as an existential wake-up call, causing individuals and communities of all backgrounds to seek a more complete paradigm, one that allows for a better understanding of not only the world around us, but also the world within and the possibility of an existence beyond this life.

People questioning their priorities, and people reeling from loss, changed. Even in the media, suddenly, the world of celebrity and gossip had taken a back seat to the world of altruism and selflessness. In my experience, for every one person who acted more selfishly, many more acted selflessly. People did whatever they could to serve others, with humanity and understanding, even if all they could do was offer a simple gesture of gratitude. It was extremely heartwarming to see people come out at 7 p.m. to cheer us on and show their appreciation. It was surreal and touching that even the police and fire crews, who had put themselves at risk for years, were standing outside hospitals and saluting those working inside.

As one of my nursing colleagues said, "[COVID-19] challenged our prevailing view of being self-focused, seeking comfort, and simply following pleasure—with little attention to the world of hardship and difficulty beyond—even though we all learn through hardship." After all, as she put it, during the first wave of the pandemic we all learned that "tomorrow is promised to no one."

FLIGHT OF LIFE

What is it about loss that brings out such beautiful human qualities—humanity, comradery, kindness, understanding, and empathy—and what is it about loss that makes us stand back and question our purpose? Why does it take a natural disaster to shake us into thinking more deeply?

The pandemic brought home the importance of seeking answers to our seemingly existential questions, as well as the relevance and importance of the scientific research we had been engaged with. This had been clear from the perspective of the scientist and doctor in me, to save the "lives and brains" of people. Now I felt it on a deep personal level from the perspective of a person who had faced the reality that my own time is limited and may end abruptly. It focused my mind on how our research may help people like me find meaning and optimize life. This is why knowing what happens with death is important for life as it helps direct what to prioritize and how to live. *Knowing the purpose directs the process.*

Imagine being on a fifteen- to twenty-hour flight from New York to Singapore. During the flight everyone is naturally fretting about what seat to sit in or what snack or meal they will be handed. They engage themselves with the immediate matters that might maximize their comfort on what may be an otherwise uncomfortable flight, by virtue of being so long. They are concerned with the quality of the food, where to stow their bag, who they are sitting next to, the entertainment, movies, and so on. Of course, a flight is finite and either way, those fifteen to twenty hours, whether comfortable or uncomfortable, will pale into insignificance compared with the reality that awaits them in the new destination. There you will be concerned with things like finding suitable accommodation, a new occupation, a new livelihood, learning to deal with cultural nuances, rules, regulations, and so on. What had preoccupied you on the plane itself (the food, snacks, and so on) will seem completely meaningless by comparison.

Now imagine that while you are on the flight, you realize there is a risk that the plane could crash on landing. How would your priorities change on the flight? What would you be focused on? Logic would dictate that everyone should prioritize doing whatever is necessary to ensure the plane will land optimally. That means finding someone to steer the plane safely, while there is still time. After all, what really matters for everyone is to arrive at the destination and pursue their life in that reality. The flight itself is nothing more than the means or the process by which to get to the destination. However, the ultimate purpose is to arrive safely and pursue a new life. Suppose your fellow passengers ignore you? They are more interested in their movie or flagging down a flight attendant for another drink. Their actions and behavior continue as if the airplane is somehow going to be their home forever. They acknowledge the plane has to land, but to them, that can all be dealt with when the time comes. For now, they are too busy with other, more pressing issues. Some are watching their favorite movie, others picking out their meal. No matter how much you try to explain that it will be too late to deal with the landing later, they ignore you. Or if they look at you, they look at you as if you have two heads. You ask yourself, "Am I the one who is illogical and crazy, or is it the other 500 or so people who have become so preoccupied with their immediate comforts that it has made them oblivious to the far more significant reality and purpose that lies ahead?"

I know this sounds extreme. Obviously, in this hypothetical situation, everyone would have prioritized the landing. But put this analogy into the context of how most people and society prioritize and deal with various matters in life—not just the question of what happens with death—and you will see it resonates. We saw the devastating consequences of putting off preparations for COVID-19 in terms of the sheer numbers of lives lost, families destroyed, and economies devastated. Even though many had warned against putting off dealing with the inevitability of the pandemic, those responsible were unprepared and uninterested. Even as China was being engulfed, many continued to convince themselves that it would just be another cold or flu. They were unwilling to plan, because they were too preoccupied with other things, until it was too late.

In my experience, when it comes to the question of what happens when we die, there are two types of people. The first are those who are completely uninterested, would rather ignore death outright and bury their head in the sand, hoping it's just never going to happen. They keep themselves preoccupied with everything else in life, until it's too late, instead of facing this question up front.

The second group of people are those who can't help but contemplate death and can't live without knowing what happens after we die. They believe there's a higher purpose to living, and they're constantly striving to gain those answers. They just can't live without them, and they feel that it is vital to know the answers because those answers will fuel the way they're going to plan and execute their life in a more meaningful way.

I am in the latter group. I categorize these people as well as myself as "the planners." We recognize the importance of planning for a different eventuality, in the same way someone in their twenties or thirties is working but is fully focused on the inevitability of their retirement and is planning for it in advance.

Considering how long the Earth has been turning, is there any real difference between a fifteen- or twenty-hour journey on a flight and a journey lasting eighty years, which is the expected length of our own flight of life? In relative terms, they are both transient flashes in time. As with the passengers on the hypothetical flight, it is natural to seek ways to optimize what may be relevant and important at different stages in life. Nobody on the plane would have been comfortable cramped between others. It is natural to try to optimize our situation as best we can. But do we become so engaged as to forget the overall purpose, or do we recognize the process for what it truly is: a means to optimize the purpose?

The main issue to try to determine in life is what is the real "signal" and what is the "noise"? For those who may be less familiar, in scientific language the signal is the meaningful information scientists are trying to detect in their studies. The noise is the unwanted variation or fluctuation arising from other, less important phenomena that interferes with the signal. To get a sense of this, imagine trying to tune in to a radio station. As you move the dial you pick up white "noise" and after a few minutes, you may pick up the real "signal" and tune in to a station.

PURPOSE

Like many of us, I have tried to detect the signal from the noise in life. Of course, there are many places for people to turn to for answers. As discussed, these are not new questions. People have faced them throughout time. The question I faced was: Where can I find answers that are compatible with a rational mind? Do we turn to others, our friends and family, society, or maybe religion, or perhaps the ever-increasing groups of people who purport to help with spiritual well-being? Or do we maybe turn to science? Where does one find help to optimize life for the signal and purpose we have talked about?

Society has its own measures of success, but it never made sense to me to limit the deeper purpose of life to these things, which, beyond a certain level of necessity, rarely aid genuine fulfilment or happiness. Wealth is a prime example. Psychologists have shown that beyond a certain threshold, getting richer is not associated with greater happiness or satisfaction; in fact, it may have a detrimental effect on well-being. Yet paradoxically, society cherishes and promotes wealth and the lives of the wealthy.

Psychologists call this phenomenon of chasing pleasure the "hedonic treadmill." You know you are on it if you are pursuing one activity after the next to feel the transient sensations of pleasure, power, or status. The problem with this pursuit, however, is once we become accustomed to a pleasant sensation, we naturally want more of it. So, in effect, we are constantly increasing what we think we need to be happy or to feel pleasure or satisfaction. Of course, what we desire can seem benign—such as tastier food, better vacations, better cars, better promotions, etc. On the face of it, each desire may seem harmless, but if these desires drive our lives and become our purpose, we are walking along that hedonic treadmill, filling our lives with limited pursuits that distract us from what may be more meaningful. Ironically, science has shown that the more self-centered we are, the less happy we are. By contrast, researchers, such as Dr. Sonja Lyubomorsky of the University of California, Riverside, and author of *The Myths of Happiness: What Should Make You Happy but Doesn't, What Shouldn't Make You Happy but Does*, have discovered that it is resiliency and growth in the face of hardship and the selfless ability to help others that lead to greater happiness and fulfilment.

The passengers ignoring the impending crash are like all of us, in denial about the ultimate fact of our life: its end. Most people keep themselves so engaged in the minutiae of how to optimize their comfort and well-being in our brief hours of life, like those passengers, that they forgo any need to think beyond. These distractions manifest as concerns about status, wealth, career, beauty, or relationships. Some of these are necessary rungs on the ladder of life, some distractions that masquerade as the goal or purpose of life and become part of the hedonic treadmill of life. Once on that treadmill, we run faster and faster toward our superficial goals of more, bigger, better, faster, all the while staying firmly in place. We may feel like we are moving forward, but we're not really going anywhere significant.

This is where the human side of the scientific study of recalled experiences of death comes in. We've discovered from thousands of testimonies that in death we learn how to live. No one comes back from the grey zone determined to upgrade their car, or buy that bigger house, or overhaul their wardrobe, or regretting missing out on a promotion. Instead, they come back with a determination to simply live with a higher purpose. In 2020 I felt that firsthand. During that first wave of the pandemic, the loudest noise from my own conscience was: *How much effort did you make to become a better human being? What have you learned from your errors? What did you accomplish as a human being?* My career was important inasmuch as it allowed me and my family to live, and to the extent that it may have been of help to others. *What had I accomplished as a father, as a son, as a husband, and so on? Had I gone out of my way to assist others?*

Whose lives had I helped to improve? Importantly, what were the opportunities where I could have done more for others, but decided not to? This was like a powerful driving thought in my mind. Many other aspects of my life seemed to pale in comparison.

There is also the reality that everyone's life and needs are different, and even the same person has different needs and priorities at different stages of life: my twenty-five-year-old self had to work and study hard to overcome many challenges in order to lay the foundations for the security of my current, fifty-year-old self. I, in turn, now need to think about what will be relevant and meaningful to my future seventy-five-year-old self. If am financially secure, then what is worthwhile and meaningful may be to spend more time helping others, without any expectation of reward, rather than getting stuck on the hedonic treadmill. Each person's life, their own calling, their personal pursuits, and what is worthwhile will probably vary depending on their life circumstances. However, the need to separate the "signal" from the "noise" and find what has meaning is critical and relative to the endgame. Given what we know is going to be the endgame in the grey zone of death, I need to engage with my life in a manner today that is going to also be optimal for the reality that I will face in death.

To answer their questions, people have traditionally turned to religion. However, there are so many contradictions among the beliefs put forth by different religions that they often leave people with many unanswered questions. Nonetheless, at the heart of most religions there is an emphasis on becoming a better human being and respecting the rights of others, even if some supposedly devout religious people don't follow these basic fundamental principles. It is also interesting that this is the only component of religions that stands out for people in death. In the grey zone, people don't focus on or highlight the ritualistic aspects of religion, or its sociocultural aspects, just the application of their core moral principles. *This is the main common unifying ground between all true religions.*

However, when it comes to helping people make sense of what happens after death, there are multiple permutations and contradictions among faiths. The beliefs are often based on stories first told thousands of years ago. Often these stories are told in a manner that would help people of a particular society and time period understand them—using depictions that were familiar to people then. However, these depictions and rituals may not resonate now. For instance, as discussed, many religions describe a hell of burning fire and despair or heaven as an everlasting beautiful place of happiness and joy. Yet during the recalled experience of death, people come to understand that such depictions are overly simplistic. Instead, they report a hierarchy of knowledge and truths, like an educational curriculum covering the different facets and levels of humanity. It is the purpose for which they may return back to life while recognizing that one life may not be sufficient to achieve all the necessary levels of knowledge. The only "hell" they come across represents the painful consequences of their own actions, and the only "heaven" also represents the consequences of their own actions.

Some people try to disembark the hedonistic treadmill by turning to contemporary spirituality. Spirituality originally meant the search for higher meaning in life, moral and ethical growth, and a desire to draw closer to "God," meaning development of ethical and moral virtues. Today, this term is increasingly used to refer to a group of diverse activities, ranging from altered states of consciousness to inner peace, harmonious relationships, music, chanting, contemplative practices (meditation, yoga, Tai Chi, and so on), psychic powers, astrology, and even the use of hallucinogenic drugs in search of so-called spiritual awakening.

However, these activities are heavily focused on inducing certain emotional states in people, not developing our humanity through rationality and reason. It is also hard to see how they can lead to greater knowledge in line with what people report with their experience of death. Nor can they clearly answer questions about purpose or meaning in life. During the recalled experience of death, none of these increasingly popular practices are highlighted or stand out as having had meaningful value either (as we saw with traditional religious ritualistic practices).

If we take what is understood to be meaningful and the signal in life based on what people come to understand during their recalled experience of death, then an "authentic" form of spirituality would have to be defined very differently from what is being labeled as "spirituality" in our times. Such an authentic spirituality should represent a universally applicable rational method, which when applied would lead a person to develop their humanity and increase their cognitive understanding of real truths, as reflected in the layers and hierarchy of knowledge and understanding that survivors experience in the grey zone.

Before I address what has resonated with me in my own journey of life, let me address what I think will be the solution for the future—namely, that science and the scientific method need to help us address our deepest existential questions, including what is authentic spirituality, in an impartial manner.

SCIENCE: NOT A FIELD OF REJECTION

Science deals with understanding universal truths, yet within the camp of "rational thinking" there is a vast void when it comes to meaning and purpose, as with our existential questions. We have seen that this is especially clear when we talk about and investigate death. Today death is discussed and studied by people in the fields of psychology, anthropology, philosophy, and theology. In Western societies, such as the United States or the United Kingdom, where I have lived, people debate and discuss life and death at a highly intellectual and philosophical level. Anthropologists seek to understand how it affects and is treated by different cultures and societies. Psychologists look at death through the lens of grief. Theologists examine it from the perspective of what gives meaning to our lives. And philosophers have been concerned with the questions: What is consciousness? Who am I? What is the self? What is the relation between the mind and brain, or the so-called mind–body problem?

By starting to look at death through the lens of science and medicine, we're able to look at things more objectively. For instance, by using natural language processing and artificial intelligence to allow us to utilize the objective and unbiased power of mathematics, we were able to determine with certainty that people's recalled experience of death is not like imaginary, hallucinatory, or dreamlike experiences. Until then, people had addressed the nature of these experiences subjectively based on their own beliefs.

Nonetheless, despite their enormous power, mainstream science and medicine have so far never seriously explored the question of death outside of a quest to live well and die comfortably. In particular, they have never engaged in the larger question of what happens when we die. In fact, every attempt is often made to downplay the significance of their discoveries to separate science from what are perceived as religious and philosophical questions.

Unlike the stereotypes about science and scientists—in my mind, science should not be a subject that rejects everything when it comes to human meaning and purpose. It would be rather irrational and unscientific to refuse to acknowledge the importance of questions that relate to us all as humans.

Science tries to answer all human questions impartially and openly. I know that some scientists—based on their own personal beliefs—may be closed to certain areas of discovery. However, an impartial scientist recognizes their own limitations and realizes that it would be shortsighted to reject things simply because the scientific tools needed to explore them may not have been discovered or because they themselves may not have yet understood them.

Newton's willingness to engage with what some now consider unorthodox areas shows the importance of being open-minded and humble. Einstein is a prime example of another scientist who humbly recognized human limitations in addressing the complexities of hidden realities.

In his book *The Evolution of Physics*, coauthored with Leopold Infeld, he explains:

> In our endeavor to understand reality, we are somewhat like a man trying to understand the mechanism of a closed watch. He sees the face and the moving hands, even hears it ticking, but he has no way of opening the case. If he is a genius, he may form some picture of a mechanism which could be responsible for all the things he observes, but he may never be quite sure his picture is the only one which could explain his observations. He will never be able to compare

> his picture with the real mechanism and he cannot even imagine the possibility of the meaning of such a comparison.

Einstein acknowledges that at any given time, scientists simply create models with which to try to explain how things work as best as possible. Importantly, those models do not represent the exact realities that they are trying to explain. They are just models. This is why the models need to be adjusted on a regular basis, since the true reality of any phenomenon they are trying to study remains hidden under ever deeper and more complex hidden layers.

One model may explain one layer of reality, but other models are needed to explain a deeper layer, and still others for deeper layers and so on. Discovering truths is like peeling back layers of an onion. Finding the first layer does not mean we have discovered the entire truth of a given matter.

For instance, we were all taught the atomic model, with a central nucleus with electrons orbiting around it. But this is not an exact model of how the atomic world really works. It is just a simplified representation to help us understand one superficial layer of its reality. Equally, a map of London isn't an exact replica of that city. It is a representation or model that is perfectly functional and allows people to get about that city. It would be a mistake to think the map represents the exact reality that is London. Likewise, our current model of the brain and consciousness can only explain the "easy" problems of consciousness and relate different conscious experiences to different parts of the brain. However, this model doesn't answer the deeper questions of how consciousness and thoughts come to exist. For that, we need a different model, one that I am convinced would view the human mind and consciousness as an undiscovered scientific entity, much like an electromagnetic field that can carry information. This might work in the same way that we send and receive fluxes of information in the form of electromagnetic waves. These waves can travel thousands of miles to be decoded by a television set.

In this model, consciousness is like a flux of energy, which can interact with the brain and is itself impacted by brain processes. However, it is not produced by the brain. In this scenario, consciousness can continue to exist beyond death, in line with our emerging scientific data. That also explains why people feel that in death, their liberated consciousness is vast, and they can suddenly understand so much more and process millions of pieces of information all at once or penetrate other people's thoughts, all while universally relating their selfhood to their consciousness, not their body. This model allows for us to take seriously the testimonies and data from hundreds of millions of survivors who have encountered the grey zone. However, when we try to stick to the old model, and assume the brain and consciousness are essentially locked together, we run into problems. That is because that model is not consistent with the realities and data we face. If, despite the lack of evidence, we continue to insist that consciousness and thoughts are produced by brain processes, then we are also forced to reject people's testimonies.

In science, these problems arise when some scientists take the models they have learned as representing all reality about a given subject. This is why they reject new discoveries or phenomena—including the recalled experience of death—that contradict their views and mental representations of a subject. It is almost as if what they have learned—the map they have created based on what had been discovered at a given time—must represent all reality about a given subject forever. This behavior is of course reinforced by the effect of mental heuristics that lead people to reject new information that counters their own beliefs, and instead seek information that supports their beliefs. Instead of adjusting their beliefs according to the reality of data, they bend reality to make it fit with their own preconceived beliefs. They mistakenly convince themselves that all these testimonies of people's recalled experiences of death can be explained by a camera trick or a drug-induced hallucination.

We are like the people Einstein described: consciousness is a closed watch that we struggle to understand and cannot conclusively answer. Even if we pooled together the combined knowledge of every great mind in the world (somewhat replicable with artificial intelligence), we would still not have a definitive answer to "What is the origin of animal or human consciousness?"

Accepting that some problems are "closed watches," together with the recognition of the limitations of our three-dimensional brain and senses as well as our mental heuristics, may help us look at new phenomena with a healthy dose of humility.

As we have seen, notable ancient thinkers such as Plato, Aristotle, Avicenna, and a whole cadre of other Greek philosophers and Greco-Islamic and Western philosophers, as well as Eastern philosophers, have sought to address these same wider questions regarding meaning and purpose in life. The need to seek the truth of this matter has pervaded throughout time and through all societies, and so it does not seem rational to ignore ancient wisdom either. The scientific method turned medicine from a hotchpotch of contradictory beliefs a couple of centuries ago into an objective and universal field of enquiry that is constantly growing and accomplishing enormous feats. Likewise, the time has come to incorporate the scientific method into understanding some of the seemingly existential questions and in particular addressing the issue of meaning and purpose and what has been referred to as authentic spirituality.

EXPANDING THE STUDY OF PSYCHE, SOUL, AND CONSCIOUSNESS

Many people may know that the term *psychology* is derived from two words: *psyche*, referring to what ancient philosophers called the soul, or self, and *ology*, which means "to study." So, in a literal sense, psychology refers to the study of the psyche, soul, consciousness, or self.

In the early days of psychology, toward the late nineteenth and early twentieth centuries, neither the seemingly intangible mind nor its study—which stood in stark contrast to other hard scientific disciplines of the time, such as engineering and physics—were considered "scientific." Yet in just a hundred years, psychology (as with medicine earlier) has grown from a primitive and obscure field to one that objectively explores the reality of the mind. The scientific method has unlocked much new knowledge about the health, development, and growth of the human and animal mind. Psychologists have demonstrated that humans are composed of a biological and a psychological entity, and while the latter is physically intangible, its universal reality can nonetheless be unearthed through objective scientific means.

However, mainstream psychology shies away from integrating the issue of meaning and authentic spirituality into its field of study. There has been some progress through the relatively new field of positive psychology, which explores some of the areas that religion and philosophy have traditionally engaged with, such as the important roles of altruism, forgiveness, and gratitude in human development and well-being. However, by and large, psychology still avoids questions that relate to meaning and the authentic spiritual needs of humans. It acts is as if these can just be avoided, or that they are not an extension of what it has already started. As with the study of death in medicine, it, too, has created an arbitrary "end" for itself, beyond which it doesn't want to venture.

Although the mindset of many people remains conditioned to regard anything related to religion or spirituality and science as essentially incompatible, current evidence contradicts that view and highlights the need for these fields to all come together. That is because they are all exploring complementary and converging aspects that relate to the universality of human lives. Today, studies suggest that up to two-thirds of scientists may believe in God. Furthermore, spirituality has already become integrated into mainstream medicine and forms an integral part of end-of-life palliative care. In addition, medicine and science are directly exploring other traditionally theological and philosophical questions. These include the nature of the self, consciousness, or soul. So why not other topics, too? It is time to acknowledge the need to merge our existential questions, including what would be considered authentic spirituality, into the objective scientific method. In this way, we may gain the results and universal areas of knowledge that we do with other subjects, including medicine.

As a result of my own scientific mindset, as well as my inherent need to follow a higher signal and purpose in life, what has resonated with me on a practical level during my own day-to-day life has been the work of Ostad Elahi (1895–1974), an influential thinker, jurist, and master musician who devoted the whole of his life to the pursuit of self-knowledge and the universal existential questions that confront humanity.* This quest led him on a journey that began in a remote village steeped in mystical tradition and ascetic practices and culminated in a life of public service in the thick of society. Yet he lived according to the same principles and values in both iterations of his life. He described his approach to spirituality as "natural," in the sense that it emphasizes the use of our reason rather than our emotions as a means of cultivating our character and discovering the mechanisms that underlie the maturation of our human consciousness, or the soul, as he called it.

Throughout his life, he sought to extract the underlying moral and ethical principles, which, when applied in daily life, can help one to develop and grow one's humanity. Rather than merely examining the core principles of the religions and universal wisdom traditions from a theoretical perspective, Elahi sought to confront them in real-life situations and in direct contact with others. This ultimately led him to a novel approach to our perennial questions, one that is centered on the rational pursuit of spirituality and existential meaning, while being compatible with the demands of contemporary life.

I am acutely aware that in some quarters of science, there is a tendency to ignore human testimonies and ancient wisdom. Yet ignoring testimonies or ancient wisdom dating back thousands of years has never seemed natural to me. Throughout my scientific and medical training, this strong dissonance between such narrow thinking by some scientists and the importance of values of ancient wisdom persisted, especially as I came to increasingly recognize the limits of our scientific models.

As you have seen, it seemed completely irrational to me that so many scientists in psychology and other fields who study consciousness choose to ignore thousands of years of insights that have led to their respective

* For more information on the life and work of Ostad Elahi, please refer to www.ostadelahi.com. For those interested in how his approach may be applied and practiced in daily life, see *Fundamentals of the Process of Spiritual Perfection: A Practical Guide* by Bahram Elahi, MD (New York: Monkfish Book Publishing Company, 2022).

fields. This is despite the fact that their work builds on and is a continuation of the work of philosophers and thinkers on the concept of the psyche, or soul.

I also recognize that there is no culture without an emphasis on morality and ethics and the deeper questions of what happens when we die, including the nature of consciousness. To me, viewing such existential questions rooted in ancient wisdom as a case of mass foolishness or a sign of being intellectually weak (as some modern scientists have tried to propose) is quite weak on its face.

Ultimately, a science of disappearance—one that considers our existence as a transient flash before eternal disappearance—is at its core meaningless. Why? Because such a science does not represent a true assertion. In other words, it doesn't represent anything that has been discovered and can be asserted to be true, which is the hallmark of the scientific method. Without the ability to assert truths, we are left with questions to explore, not reject. This is why, as mentioned previously, after our AWARE-I and -II studies, we concluded that "although systematic studies have not been able to absolutely prove the reality or meaning of patients' experiences and claims of awareness in relation to death, it has been impossible to disclaim them either. The recalled experience surrounding death now merits further genuine empirical investigation without prejudice."

The scientific persistence of seeking and pursuing answers to these perennial questions and the universal interest in them itself highlight their importance and significance. Perhaps in this context, the study of the passage to death—the grey zone—is a worthy scientific and philosophical pursuit, even though in our work we are able to study only people who come back, since we cannot access those who never made it back.

Nonetheless, in view of what has been discovered, together with the fact that we don't have any tools to explore the metaphysical level, our work in the grey zone of death has presented interesting questions and intriguing opportunities for further exploration.

We know that these discoveries don't completely resolve the scientific questions of consciousness, but they do help bring newer insights into this important field by showing that the old-fashioned model that asserts our selfhood is simply a by-product of brain processes and is annihilated with death cannot be the answer.

Placing a magnifying glass on what happens in the grey zone of death—during our transition from life to death—has started to inform us about the intriguing biology of life and death medically and has also brought a new perspective and raised more questions regarding the experiencing brain. In this manner, the study of the underlying process raises even more questions because beyond the subjective features that people recall, it is also related to the objective truth of what humans experience through the disinhibition of brain processes, which we have now found paradoxically unleashes the full capacity of other parts of the brain and human consciousness.

Ultimately, as my own personal, philosophical, and scientific knowledge has expanded, I have remained a student of the very edge of existence, by which I mean the edge of life and death, where I have been pushing the boundaries ever further.

LOOKING TO THE FUTURE

I have no doubt that all the incredible discoveries that I have pointed out in this book so far represent just a very small tip of a much larger field of discovery. We are at the cusp of the exploration of a new frontier in science. As I look forward, I am excited to think about what will be discovered. I have little doubt that, in the future, people who would be declared dead today will be routinely brought back to life, just as the use of CPR to revive people today would have been viewed as fantasy a hundred years ago. I have little doubt that the exciting work that Dr. Montgomery and others have pioneered will enable us to make organs available to everybody who needs them. I have little doubt that it will be possible to grow full human brains outside of the body and to replace diseased parts of the brain. I am also quite optimistic that the nature of human consciousness—our very selfhood—will be discovered to be a flux of energy like electromagnetic waves that interacts with the brain and body but is not produced by them.

However, from where I stand right now, I will not be alive to see many of these advances. In the same way that so many people in my own family will have left this life in the next few decades, I, too, will be on the same train. An online life expectancy calculator, available on the United States Social Security Administration website, estimates that I have exactly thirty years from now until my departure. Another calculator gives me 632 months of life left and predicts I will die on Saturday, October 14, 2051. Only time will tell whether these calculations are going to be accurate in my case.

What has become clear to me from my research is that what matters the most during our lives is the development and growth of our humanity. The real question, then, is: How should I live my life today in a way to accomplish this objective? It is within this context that I would seek to define the meaning and purpose of our lives and the role of spirituality.

I know I'm not the only person contemplating these questions, but I hope that by writing this book, I will inspire others to help answer them.

ACKNOWLEDGMENTS

THROUGHOUT THIS BOOK, I HAVE SOUGHT TO CONVEY THE FASCINATING scientific story that is emerging at the interface of life, death, and beyond. Nothing is ever carried out alone, and this work is no different, as it represents the efforts and input from hundreds of people to whom I am forever indebted.

First, I would like to express my deepest and most sincere gratitude to Professor Ebby Elahi for his time, discussions, and unique insights over the past two decades—starting from when I was writing my first book, *What Happens When We Die*, until now. Many of the key ideas presented in this book and elsewhere, including the role of disinhibition, have came about directly from those discussions.

I must also express my extreme gratitude to Dr. Peter Fenwick for his kind and generous support in the early years of my career, as well as Dr. Derek Waller, who enabled us to start the research at Southampton General Hospital in England. I will always be thankful to Heather Sloan for the time she dedicated to my studies. These were the people who enabled the foundations of this work to be laid down more than two decades ago. More recently, numerous colleagues and research staff at multiple hospitals have helped us continue that work.

I would like to thank Mary Curran, who over the course of almost five years helped me with the writing of this book, as well as Caroline Greeven, for her tremendously valuable input during the completion of this book. I am also greatly appreciative of Samantha Olson, who helped research and prepare the first draft of chapter 1. Her input and enthusiasm will be forever etched in that story.

A huge thank-you also goes to Andrew Stuart, my agent, and Lauren Marino, my editor at Hachette, for their patience and guidance throughout the entire process.

I owe a tremendous and sincere thanks to my wife, who has lived through and supported me with my work, as well as my mother and daughter. Without their sacrifices, I could not have carried out my research.

Last but not least, I am deeply grateful to all of the doctors and scientists who provided their outstanding insights regarding the scientific study of life and death, as well as all the individuals who were willing to share their profound stories about their own recalled experiences of death throughout this book.

BIBLIOGRAPHY AND SUGGESTED FURTHER READING

For those who wish to explore some of the key concepts and scientific studies that have been discussed (but not already directly referenced in the book), the following materials and notes are provided:

SECTION ONE: SCIENCE OF LIFE AND DEATH

Chapter 1: The Brain in the Bucket

The meeting held at the NIH on March 28, 2018, remains publicly available at the time of writing this book and can be accessed at:

NIH BRAIN Initiative Workshop on Research with Human Neural Tissue, NIH VideoCast, https://videocast.nih.gov/watch=27227.

Other key publications are:

Vrselja, Z., S. G. Daniele, J. Silbereis, et al., "Restoration of brain circulation and cellular functions hours post mortem," *Nature* 568 (2019): 336–343.

Palmer, T. D., P. H. Schwartz, P. Taupin, B. Kaspar, S. A. Stein, and F. H. Gage, "Progenitor cells from human brain after death," *Nature* 411 (2001): 42–43.

Marfia, G., L. Madaschi, F. Marra, et al., "Adult neural precursors isolated from post mortem brain yield mostly neurons: An erythropoietin dependent process," *Neurobiology of Disease* 43 (2011): 86–98. doi: 10.1016/j.nbd.2011.02.004.

Abbas, F., S. Becker, B. W. Jones, et al., "Revival of light signalling in the postmortem muse and human retina," *Nature* 606, no. 7913 (2022): 351–357.

Youngner, S., and Hyun I., "Pig experiment challenges assumptions around brain damage in people," *Nature* 568 (2019): 302–304.

Andrijevic, D., Z. Vrselja, T. Lysyy, et al., "Cellular recovery after prolonged warm ischemia of the whole body," *Nature* 608 (2022): 405–412.

Onorati, M., L. Zhen, L. Fuchen, et al., "Zika virus disrupts phospho-TBK1 localization and mitosis in human neuroepithelial stem cells and radial glia," *Cell Reports* 16 (2016): 2576–2592.

This is the scientific article in which Dr. Sestan and his team highlighted their ability to grow brain cells from a donated human brain received forty-eight hours after death (due to an error by the courier company). This information can be found in the supplementary section in Table S3). There is a sample code (e.g., HSB 339), followed by a postmortem interval (PMI) in hours (e.g., 24.72 hours), which refers to how long after death the brain had been collected for the study. Samples HSB 341 and 343 were from brains obtained 48 and 49.5 hours after death. Specifically, it says:

HSB 339 **24.72 hr** PMI
HSB 341 **48 hr** PMI
HSB 343 **49.5 hr** PMI
HSB 344 **22.5 hr** PMI

Chapter 2: A New Scientific Frontier

Cooper, J. A., J. D. Cooper, and J. M. Cooper, "Cardiopulmonary resuscitation: History, current practice, and future direction," *Circulation* 114 (2006): 2839–2849.

Paradis, N. A., H. R. Halperin, K. B. Kern, V. Wenzel, and D. A. Chamberlain (eds.), "Preface," in *Cardiac Arrest: The Science and Practice of Resuscitation Medicine*. New York: Cambridge University Press, 2007.

Eisenberg, M. S., P. Baskett, and D. Chamberlain, "A history of cardiopulmonary resuscitation," in Paradis, N. A., H. R. Halperin, K. B. Kern, V. Wenzel, and D. A. Chamberlain (eds.), *Cardiac Arrest: The Science and Practice of Resuscitation Medicine*. New York: Cambridge University Press, 2007, pp. 3–25.

Kouwenhoven, W. B., J. R. Jude, and G. G. Knickerbocker, "Closed-chest cardiac massage," *JAMA* 173 (1960): 1064–1067.

Wijdicks, E. F., *Brain Death*, 2nd ed. New York: Oxford University Press, 2011.

Seifi, A., J. V. Lacci, D. A. Godoy, "Incidence of brain death in the United States," *Clinical Neurology and Neurosurgery* 195 (2020): 105885.

Neumar, R. W., J. P. Nolan, J, C. Adrie, et al., "Post-cardiac arrest syndrome: Epidemiology, pathophysiology, treatment, and prognostication. A consensus statement from the International Liaison Committee on Resuscitation (American Heart Association, Australian and New Zealand Council on Resuscitation, European Resuscitation Council, Heart and Stroke Foundation of Canada, InterAmerican Heart Foundation, Resuscitation Council of Asia, and the Resuscitation Council of Southern Africa); the American Heart

Association Emergency Cardiovascular Care Committee; the Council on Cardiovascular Surgery and Anesthesia; the Council on Cardiopulmonary, Perioperative, and Critical Care; the Council on Clinical Cardiology; and the Stroke Council," *Circulation* 118 (2008): 2452–2483.

Rubenstein, A., E. Cohen, and E. Jackson, "The Definition of Death and the Ethics of Organ Procurement from the Deceased," *The President's Council on Bioethics*, 2006, https://bioethicsarchive.georgetown.edu/pcbe/background/rubenstein.html.

Burkle, C. M., S. Am, and E. F. Wijdicks, "Brain death and the courts," *Neurology* 76 (2011): 837–841.

Wijdicks, E. F., and E. Pfeifer, "Neuropathology of brain death in the modern transplant era," *Neurology* 70 (2008): 1234–1237.

Lewis, A. "The Uniform Determination of Death Act is being revised," *Neurocritical Care* 36 (2022): 335–338.

Chapter 3: Exploring Death: Past to Present

Moody, R. A., *Life After Life*. New York: Bantam Press, 1975.

Parnia, S., S. G. Post, M. T. Lee, et al., "Guidelines and standards for the study of death and recalled experiences of death—a multidisciplinary consensus statement and proposed future directions," *Annals of the New York Academy of Sciences* 1511 (2022): 5–21.

For a full discussion of the different studies that have been mislabeled as "near-death experiences," please refer to the supplementary sections, including the supplementary tables.

For specific studies, please refer to the references in the supplementary materials as well as the references in the main manuscript. Some of these studies (which are also referenced in the consensus statement already and, at the time of writing this book, remain available to the public and professionals alike) are also listed below.

Other publications that have expressed a variety of opinions on so-called near-death experiences:

Fenwick, P., and E. Fenwick, *The Truth in the Light: An Investigation of over 300 Near-Death Experiences*. London: Hodder Headline, 1995.

Sabom, M., *Recollections of Death: A Medical Investigation*. New York: Harper & Row, 1983.

Ring, K., *Life at Death*. New York: Coward McCann, 1980.

Khoshab, H., S. Seyedbagheri, S. Iranmanesh, et al., "Near-death experience among Muslim cardiopulmonary resuscitation survivors," *Iranian Journal of Nursing and Midwifery Research* 25 (2020): 414–418.

Pascricha, S., and I. Stevenson, "Near-death experiences in India," *Journal of Nervous Mental Disease* 55 (1986): 542–549.

Feng, Z., "A research on near-death experiences of survivors in big earthquake of Tangshan 1976," *Zhonghua Shen Jing Jing Shen Ke Za Zhi* [Chinese Journal of Neurology and Psychiatry] 25 (1992): 222–225, 253–254.

Gallup, G., and W. Proctor, *Adventures in Immortality: A Look Beyond the Threshold of Death*. New York: McGraw-Hill, 1982.

Perera, M., G. Padmasekara, and J. Belanti, "Prevalence of near-death experiences in Australia," *Journal of Near-Death Studies* 24 (2005): 109–116.

Nelson, K. R., M. Mattingly, S. A. Lee, and F. A. Schmitt, "Does the arousal system contribute to near-death experiences?" *Neurology* 66 (2006): 1003–1009.

Beauregard, M., J. Courtemanche, and V. Paquette, "Brain activity in near-death experiences during a meditative state," *Resuscitation* 80 (2009): 1006–1010.

Charland-Verville, V., J.-P. Jourdan, M. Thonnard, et al., "Near-death experiences in non-life-threatening events and coma of different etiologies," *Frontiers in Human Neuroscience* 8 (2014): 203.

Lempert, T., "Syncope and near-death experience," *Lancet* 344 (1994): 829–830.

Timmerman, C., L. Roseman, D. Williams, et al., "DMT models the near-death experience," *Frontiers in Psychology* 9 (2018): 1424.

Owens, J. E., E. W. Cook, and I. Stevenson, "Features of 'near death experience' in relation to whether or not patients were near death," *Lancet* 336 (1990): 1175–1177.

Kondziella, D., J. Dreier, and M. H. Olsen, "Prevalence of near-death experiences in people with and without REM sleep intrusion," *PeerJ* 7 (2019): e7585.

Devinsky, O., and G. Lai, "Spirituality and religion in epilepsy," *Epilepsy and Behavior* 12 (2008): 636–643.

Britton, W. B., and R. R. Bootzin, "Near-death experiences and the temporal lobe," *Psychological Science* 15 (2004): 254–258.

MacLullich, A. M. J., K. J. Ferguson, T. Miller, et al., "Unravelling the pathophysiology of delirium: A focus on the role of aberrant stress responses," *Journal of Psychosomatic Research* 65 (2008): 229–238.

Whinnery, J. E., "Psychophysiologic correlates of unconsciousness and near-death experiences," *Journal of Near-Death Studies* 15 (1997): 231–258.

Greyson, B., "The Near-Death Experience Scale: Construction, reliability, and validity," *Journal of Nervous and Mental Disease* 171 (1983): 369–375.

Martial, C., J. Simon, N. Puttaert, et al., "The Near-Death Experience Content (NDE-C) scale: Development and psychometric validation," *Consciousness and Cognition* 86 (2020): 103049.

Martial, C., H. Cassol, V. Charland-Verville, C. Pallavicini, C. Sanz, and F. Zamberlan, "Neurochemical models of near-death experiences: A large-scale study based on the semantic similarity of written reports," *Consciousness and Cognition* 69 (2019): 52–69.

Blackmore, S. J., and T. Troscianko, "The physiology of the tunnel," *Journal of Near-Death Studies* 8 (1988): 15–28.

Blackmore, S. J., "Near-Death Experiences," *Journal of the Royal Society of Medicine* 89 (1996): 73–76.

Whinnery, J. E., "Psychophysiologic correlates of unconsciousness and near-death experiences," *Journal of Near-Death Studies* 15 (1997): 231–258.

Klemenc-Ketis, Z., S. Grmec, and J. Kersnik, "The effect of carbon dioxide on near-death experiences in out-of-hospital cardiac arrest survivors: A prospective observational study," *Critical Care* 14 (2010): R56.

Carr, D. B., "Endorphins at the approach of death," *Lancet* 1 (1981): 390.

Sotelo, J., R. Perez, P. Guevara, and A. Fernandez, "Changes in brain, plasma and cerebrospinal fluid contents of B-endorphin in dogs at the moment of death," *Neurological Research* 17 (1995): 223–225.

Morse, M., D. Venecia, and J. Milstein, "Near-death experiences: A neurophysiologic explanatory model," *Journal of Near-Death Studies* 8 (1989): 45–53.

Jansen, K., "Near-death experience and the NMDA receptor," *British Medical Journal* 298 (1989): 1708.

Appleton, R. E., "Reflex anoxic seizures," *British Medical Journal* 307 (1993): 214–215.

Carr, D., "Pathophysiology of stress-induced limbic lobe dysfunction: A hypothesis for NDEs," *Journal of Near-Death Studies* 2 (1982): 75–89.

Parnia, S., D. Waller, R. Yeates, and P. Fenwick, "A qualitative and quantitative study of the incidence, features and aetiology of near-death experiences in cardiac arrest survivors," *Resuscitation* 48 (2001): 149–156.

Sheak, K. R., D. J. Wiebe, M. Leary, et al., "Quantitative relationship between end-tidal carbon dioxide and CPR quality during both in-hospital and out-of-hospital cardiac arrest," *Resuscitation* 89 (2015): 149–154.

Greyson, B., and J. P. Long, "Does the arousal system contribute to near-death experience?" *Neurology* 67 (2006): 2265.

Husain, A. M., P. P. Miller, and S. T. Carwile, "REM sleep behavior disorder: Potential relationships to post-traumatic stress disorder," *Journal of Clinical Neurophysiology* 18 (2001): 148–157.

For a discussion on brain organoids:

Please refer to the publicly available meeting held at the NIH on March 28, 2018, which is also under the recommended materials for chapter 1. https://videocast.nih.gov/watch=27227.

Chapter 4: Inside the Dying Brain: Bursts of Activity

Xu, G., T. Mihaylova, D. Li, et al., "Surge of neurophysiological coupling and connectivity of gamma oscillations in the dying human brain," *Proceedings of the National Academy of Sciences of the United States of America* 120, no. 19 (2023): e2216268120.

Borjigin, J., U. Lee, T. Liu, et al., "Surge of neurophysiological coherence and connectivity in the dying brain," *Proceedings of the National Academy of Sciences of the United States of America* 110 (2013): 14432–14437.

Chawla, L. S., S. Akst, C. Junker, B. Jacobs, and M. G. Seneff, "Surges of electroencephalogram activity at the time of death: A case series," *Journal of Palliative Medicine* 12 (2009): 1095–1100.

Norton, L., R. M. Gibson, T. Gofton, et al., "Electroencephalographic recordings during withdrawal of life-sustaining therapy until 30 minutes after declaration of death," *Canadian Journal of Neurological Sciences* 44 (2017): 139–145.

Matory, A. L., A. Alkhachroum, W. T. Chiu, et al., "Electrocerebral signature of cardiac death," *Neurocritical Care* 35 (2021): 853–861.

Vicente, R., M. Rizzuto, C. Sarica, et al., "Enhanced interplay of neuronal coherence and coupling in the dying human brain," *Frontiers in Aging Neuroscience* 14 (2022): 813531.

Pani, P., F. Giarrocco, M. Giamundo, et al., "Persistence of cortical neuronal activity in the dying brain," *Resuscitation* 130 (2018): e5–e7.

Herrmann, N., "What is the function of the various brainwaves?" *Scientific American*, December 22, 1997.

Parnia, S., T. Keshavarz Shirazi, J. Patel, et al., "AWAreness during REsuscitation – II: A multi-center study of consciousness and awareness in cardiac arrest," *Resuscitation* 191 (2023): 109903.

Parnia, S., D. Waller, R. Yeates, and P. Fenwick, "A qualitative and quantitative study of the incidence, features and aetiology of near-death experiences in cardiac arrest survivors," *Resuscitation* 48 (2001): 149–156.

Parnia, S., K. Spearpoint, G. de Vos, et al., "AWARE-AWAreness during Resuscitation—A prospective study," *Resuscitation* 85 (2014): 1799–1805. doi: 10.1016/j.resuscitation.2014.09.004.

van Lommel, P., R. van Wees, V. Meyers, and I. Elfferich, "Near-death experience in survivors of cardiac arrest: A prospective study in the Netherlands," *Lancet* 358 (2001): 2039–2045.

Chapter 5: Mystery No More: Disinhibition Not Degeneration

Some studies that highlight how some genes become activated after death, as well as those that refer to autoresuscitation and the brain's response to the cessation of the heartbeat, are provided. These are not meant to be exhaustive, but are meant to provide some directions for further reading.

Scott, L., S. Finley, C. Watson, et al., "Life and death: A systematic comparison of antemortem and postmortem gene expression," *Gene* 731 (2020): 144349.

Dachet, F., J. B. Brown, T. Valyi-Nagy, et al., "Selective time-dependent changes in activity and cell-specific gene expression in human postmortem brain," *Scientific Reports* 11, no. 1 (2021): 6078.

Ferreira, P. G., M. Muñoz-Aguirre, F. Reverter, et al., "The effects of death and post-mortem cold ischemia on human tissue transcriptomes," *Nature Communications* 9, no. 1 (2018): 490.

Pozhitkov, A. E., R. Neme, T. Domazet-Lošo, B. G. Leroux, S. Soni, D. Tautz, and P. A. Noble, "Tracing the dynamics of gene transcripts after organismal death," *Open Biology* 7 (2017): 160267.

Abouhashem, A. S., K. Singh, R. Srivastava, et al., "The prolonged terminal phase of human life induces survival response in the skin transcriptome," bioRxiv 2023.05.15.540715. doi: 10.1101/2023.05.15.540715 (at the time of publication, this study, by Dr. Chandan K. Sen and his team, was still undergoing peer review).

Gordon, L., M. Pasquier, H. Brugger, et al., "Autoresuscitation (Lazarus phenomenon) after termination of cardiopulmonary resuscitation—a scoping review," *Scandinavian Journal of Trauma, Resuscitation and Emergency Medicine* 28 (2020): 14.

Hornby, K., L. Hornby, S. D. Shemie, "A systematic review of autoresuscitation after cardiac arrest," *Critical Care Medicine* 38 (2010): 1246–1253.

Dunser, M. W., S. Jochberger, K.-H. Stadlbauer, and V. Wenzel, "The neuroendocrine response to global ischemia and reperfusion," in Paradis, N. A., H. R. Halperin, K. B. Kern, V. Wenzel, and D. A. Chamberlain (eds.), *Cardiac Arrest: The Science and Practice of Resuscitation Medicine*. New York: Cambridge University Press, 2007, pp. 128–162.

Chapter 6: The Lucid Death

Parnia, S., S. G. Post, M. T. Lee, et al., "Guidelines and standards for the study of death and recalled experiences of death—a multidisciplinary consensus statement and proposed future directions," *Annals of the New York Academy of Sciences* 1511 (2022): 5–21.

The description of the experience is derived from the testimonies and themes identified and summarized in this scientific publication.

SECTION TWO: EXPERIENCING THE GREY ZONE OF DEATH

The themes explored in chapters 7 through 12 are mostly discussed in the following two publications:

Parnia, S., S. G. Post, M. T. Lee, et al., "Guidelines and standards for the study of death and recalled experiences of death—a multidisciplinary consensus statement and proposed future directions," *Annals of the New York Academy of Sciences* 1511 (2022): 5–21.

Parnia, S., T. Keshavarz Shirazi, J. Patel, et al., "AWAreness during REsuscitation—II: A multi-center study of consciousness and awareness in cardiac arrest," *Resuscitation* 191 (2023): 109903.

Some of the results of the grounded theory–based analyses can be found in the supplementary materials in these two publications. In addition:

Chapter 8: Awestruck: The Doctors' Perspective

Other examples of children's experiences can be found in the following publications:

Morse M. L., "Near-death experiences of children," *Journal of Pediatric Oncological Nursing* 11 (1994): 139–144.

Serdahely, W. J., "Pediatric near-death experiences," *Journal of Near-Death Studies* 9 (1990): 33–39.

Herzog, D. B., and J. T. Herrin, "Near-death experiences in the very young," *Critical Care Medicine* 13 (1995): 1074–1075.

Morse, M., P. Castillo, D. Venecia, J. Milstein, and D. C. Tyler, "Childhood near-death experiences," *American Journal of Diseases in Children* 140, no. 11 (1986): 1110–1114.

Shears, D., S. Elison, M. E. Garralda, and S. Nadel, "Near-Death Experiences with Meningococcal Disease," *Journal of the American Academy of Child and Adolescent Psychiatry* 7 (2005): 630–631.

SECTION THREE: EXPLORING CONSCIOUSNESS

There is significant overlap between the references and topics covered in chapters 13 and 14, and so they should ideally be examined together.

Chapter 13: A Nobel Prize–Winning Idea

For a synopsis of Sir John Eccles's work that led to the Nobel Prize, please refer to:

"Sir John Eccles—Biographical," NobelPrize.org. www.nobelprize.org/prizes/medicine/1963/eccles/biographical.

Dr. Megan Craig's article published in the *New York Times* can be accessed at: www.nytimes.com/2019/06/25/opinion/brain-injury-philosophy.html.

Some of the other studies that relate to the topics discussed are:

Feuillet, L., H. Dufour, and J. Pelletier, "Brain of a white-collar worker," *Lancet* 370 (2007): 262.

Seth, A. K., and T. Bayne, "Theories of consciousness," *Nature Reviews Neuroscience* 23 (2022): 439–452.

Hameroff, S., A. Nip, M. Porter, and J. Tuszynski, "Conduction pathways in microtubules, biological quantum computation, and consciousness," *Biosystems* 64 (2002): 149–168.

Penrose, R., *Shadows of the Mind*. New York: Oxford University Press, 1994.

Penrose, R., "Consciousness, the brain, and spacetime geometry: An addendum. Some new developments on the Orch OR model for consciousness," *Annals of the New York Academy of Sciences* 929 (2002): 105–110.

Beck, F., and J. C. Eccles, "Quantum aspects of brain activity and the role of consciousness," *Proceedings of the National Academy of Sciences of the United States of America* 23 (1992): 11357–11361.

Blackmore, S., *Consciousness: An Introduction*. London: Hodder and Stoughton, 2003.

Chalmers, D. J., "The puzzle of conscious experience, mysteries of the mind," in "Mysteries of the Mind," special issue, *Scientific American* (1997): 30–37.

Rees, G., G. Kreiman, and C. Koch, "Neural correlates of consciousness in humans," *Nature Reviews Neuroscience* 4 (2002): 261–270.

Searle, J., "Do we understand consciousness?" *Journal of Consciousness Studies* 5–6 (1998): 718–733.

Tononi, G., and G. M. Edelman, "Consciousness and complexity," *Science* 282 (1998): 1846–1851.

Greenfield, S., "Mind, brain and consciousness," *British Journal of Psychiatry* 181 (2002): 91–93.

Editorial, in "Mysteries of the Mind," special issue, *Scientific American* (1997): 3.

Seth, A. K., and T. Bayne, "Theories of consciousness," *Nature Reviews Neuroscience* 23 (2022): 439–452.

Friedman, G., K. W. Turk, and A. E. Budson, "The current of consciousness: Neural correlates and clinical aspects," *Current Neurology and Neuroscience Reports* 23 (2023): 345–352.

Calabro, R. S., A. Cacciola, P. Bramanti, and D. Milardi, "Neural correlates of consciousness: What we know and what we have to learn," *Neurological Sciences* 34 (2015): 505–513.

Paulson, S., D. D. Hoffman, and S. O'Sullivan, "Reality is not as it seems," *Annals of the New York Academy of Sciences* 1458 (2019): 65–69.

Hoffman, D. D., and C. Prakash, "Objects of consciousness," *Frontiers in Psychology* 5 (2014): 577.

Chapter 14: The Self and Its Brain

Adamson, P., and F. Benevich, "The thought experimental method: Avicenna's flying man argument," *Journal of the American Philosophical Association* 4 (2018): 147–164.

Arbman, E., *Ecstasy, or Religious Trance, vol. 1, Vision and Ecstasy*. Stockholm: Scandinavian University Books, 1963.

Aristotle, *De Anima*, trans. Hugh Lawson. New York: Penguin Classics Paperback, 1986, p. 11.

Baig, M. N., F. Chishty, P. Immesoete, and C. S. Karas, "The Eastern heart and Galen's ventricle: A historical review of the purpose of the brain," *Neurosurgical Focus* 23, no. 1 (2007).

Beck, F., and J. C. Eccles, "Quantum aspects of brain activity and the role of consciousness," *Proceedings of the National Academy of Sciences of the United States of America* 89, no. 23 (1992): 11357–11361.

Bennett, M. R., "Development of the concept of mind," *Australian and New Zealand Journal of Psychiatry* 41, no. 12 (2007): 943–956.

Bennett, M. R., and P. M. S. Hacker, *Philosophical Foundations of Neuroscience*, 2nd ed. Hoboken, NJ: Wiley-Blackwell, 2022.

Blackmore, S., *Consciousness: An Introduction*. London: Hodder and Stoughton, 2003.

Bremmer, J., *The Early Greek Concept of the Soul*. Princeton, NJ: Princeton University Press, 1983.

Chalmers, D. J., "The puzzle of conscious experience, mysteries of the mind," in "Mysteries of the Mind," special issue, *Scientific American* (1997): 30–37.

Claus, D. B., *Toward the Soul: An Inquiry into the Meaning of [Psychē] Before Plato*. New Haven, CT: Yale University Press, 1981.

Collinson, D. *Fifty Major Philosophers—A Reference Guide*. Routledge, 1987.

Crick, F., and C. Koch, "Consciousness and neuroscience," *Cerebral Cortex* 8, no. 2 (1998): 97–107.

Crivellato, E., and D. Ribatti, "Soul, mind, brain: Greek philosophy and the birth of neuroscience," *Brain Res Bull* 71, no. 4 (2007): 327–336.

Demertzi, A., C. Liew, D. Ledoux, M. A. Bruno, M. Sharpe, S. Laureys, and A. Zeman, "Dualism persists in the science of mind," *Annals of the New York Academy of Sciences* 1157 (2009): 1–9.

Dennett, D., "Are we explaining consciousness yet?" *Cognition* 79 (Apr. 2001): 1–2, 221–237.

"Does neuroscience threaten human values?" *Nature Neuroscience* 1, no. 7 (1998): 535–536.

Dolan, B., "Soul searching: A brief history of the mind/body debate in the neurosciences," *Neurosurgical Focus* 23, no. 1 (2007).

Eccles, J. C., "Evolution of consciousness," *Proceedings of the National Academy Sciences of the United States of America* 89, no. 16 (1992): 7320–7324.

Editorial, in "Mysteries of the Mind," special issue, *Scientific American* (1997): 3.

Elahi, B., *Spirituality Is a Science*. New York: Cornwall Books, 1999, study 2.

Elahi, B., *Medicine of the Soul*. New York: Cornwall Books, 2001, studies 4 and 6.

Elahi, B., *The Path of Perfection*. Bracey, VA: Paraview, 2005, chapters 5–8.

Elahi, B., *The Nature of the Self*, University of Sorbonne, Paris, March 2011 (video excerpts available at www.e-ostadelahi.com [lectures]).

Elahi, B., *Fundamentals of the Process of Spiritual Perfection: A Practical Guide*. Rhinebeck, NY: Monkfish Book Publishing, 2022.

Encyclopedia Britannica, [edition] ed., s.v. "the soul."

Fenwick, P., "Current methods of investigation in neuroscience," in M. Velmans (ed.), *Investigating Phenomenal Consciousness*. Amsterdam: John Benjamins, 2000.

Flohr, H., "An information processing theory of anaesthesia," *Neuropsychologia* 33, no. 9 (1995): 1169–1180.

Flohr, H., U. Glade, and D. Motzko, "The role of the NMDA synapse in general anesthesia," *Toxicology Letters* 100–101 (1998): 23–29.

Frackowiak, R., K. Friston, C. Frith, R. Dolan, and J. C. Mazziotta (eds.), *Human Brain Function*. London: Academic Press, 1997.

Freeman, W., "Consciousness, intentionality and causality," *Journal of Consciousness Studies* 6 nos. 11–12 (1999): 143–172.

Greenfield, S., "Mind, brain and consciousness," *British Journal of Psychiatry* 181, no. 2 (2002): 91–93.

Hameroff, S., A. Nip, M. Porter, and J. Tuszynski, "Conduction pathways in microtubules, biological quantum computation, and consciousness," *Biosystems* 64, nos. 1–3 (2002): 149–168.

Henslin, J., *Down to Earth Sociology*. New York: Free Press, 2007, pp. 277–287.

Penrose, R., *Shadows of the Mind*. New York: Oxford University Press, 1994.

Penrose, R., "Consciousness, the brain, and spacetime geometry: An addendum. Some new developments on the Orch OR model for consciousness," *Annals of the New York Academy of Sciences* 929 (2001): 105–110.

Rees, G., G. Kreiman, and C. Koch, "Neural correlates of consciousness in humans," *Nature Reviews Neuroscience* 3, no. 4 (2002): 261–270.

Santoro, G., M. D. Wood, L. Merlo, G. P. Anastasi, F. Tomasello, and A. Germano, "The anatomic location of the soul from the heart, through the brain, to the whole body, and beyond: A journey through Western history, science, and philosophy," *Neurosurgery* 65, no. 4 (2009): 633–643; discussion: 643.

Searle, J., "Do we understand consciousness?" *Journal of Consciousness Studies* 5, nos. 5–6 (1998): 718–733.

Tononi, G., and G. M. Edelman, "Consciousness and complexity," *Science* 282, no. 5395 (1998): 1846–1851.

Additional notes for Chapter 14:

According to the *Stanford Encyclopedia of Philosophy*, Democritus was one of the two founders of ancient atomist theory. The atomists held that there are smallest indivisible bodies from which everything else is composed (see *Stanford Encyclopedia of Philosophy*—Democritus).

Epicureanism—A Dictionary of Philosophy, edited by Antony Flew, Pan Books, London, 1979.

For more information, please refer to *Stanford Encyclopedia of Philosophy*, [edition] ed., s.v. "Plato."

According to the *Stanford Encyclopedia of Philosophy*, for Plato "the world that appears to our senses is in some way defective and filled with error, but there is a more real and perfect realm, populated by entities (called "forms" or "ideas") that are eternal, changeless, and in some sense paradigmatic for the structure and character of our world. Among the most important of these are goodness, beauty, equality, unity, and being. These terms—"goodness," "beauty," and so on—are often capitalized by those who write about Plato, in order to call attention to their exalted status; similarly for "forms" and "ideas." The most fundamental distinction in Plato's philosophy is between the many observable objects that appear beautiful (good, just, unified, equal) and the "one" object that is what beauty (goodness, justice, unity) really is, from which those many beautiful (good, just, unified, equal, big) things receive their names and their corresponding characteristics. In many of Plato's works readers are urged to transform their values by taking to heart the greater reality of the forms and the defectiveness of the corporeal world and to recognize that "the soul is a different sort of object from the body—so much so that it does not depend on the existence of the body for its functioning, and can in fact grasp the nature of the forms far more easily when it is not encumbered by its attachment to anything corporeal."

Taking the example of a horse, then, for Aristotle, the "soul" arises from the physical characteristics of the matter that constitutes the horse—hence they are inseparable but nevertheless can be distinguished from each other (see Aristotle, *De Anima* [translated by Hugh Lawson] Penguin Classics Paperback, 1986—Introduction).

According to research by Ernst Arbman,* early beliefs in the soul involved a form of duality between a "free soul" and a "body soul." The "free soul" was believed to be active during unconsciousness and sleep and conferred individuality to a person, whereas the "body soul" conferred life and was active during states of wakefulness and consciousness. During his research, Arbman further concluded that the "body soul" was itself thought of in terms of both a "life soul"—the breath of life principle—and an "ego soul" that was responsible for psychological functions. Interestingly, at this time the eschatological† and psychological attributes of the soul were considered separately and had not yet merged into one.

A discussion on *cardiocentrism* versus *encephalocentrism* can be found in Crivellato, E., and D. Ribatti, "Soul, mind, brain: Greek philosophy and the birth of neuroscience," *Brain Research Bulletin* 71, no. 4 (2007): 327–336.

SECTION FOUR: A WORLD OF DISTORTIONS

Chapter 15: Putting Lipstick on a Pig

Dening, T., and G. E. Berrios, "Autoscopic Phenomena," *British Journal of Psychiatry* 165 (1994): 808–817.

Zamboni, G., C. Budriesi, and P. Nichelli, "'Seeing oneself': A case of autoscopy," *Neurocase* 11 (2005): 212–215.

Anzellotti, F., V. Onofrj, V. Maruotti, et al., "Autoscopic Phenomena: Case Reports and Review of Literature," *Behavioral and Brain Functions* 7 (2011): 2.

Blanke, O., S. Ortigue, T. Landis, and M. Seeck, "Stimulating illusory own-body perceptions," *Nature* 419 (2002): 269–270.

* A twentieth-century Swedish scholar who studied prevalent beliefs in the soul during early civilizations (see Arbman, E., *Ecstasy, or Religious Trance: Vision and Ecstasy,* (Stockholm: Bokförlaget, 1963), and Bremmer, J., *The Early Greek Concept of the Soul* (Princeton, NJ: Princeton University Press, 1983).

† The branch of theology concerned with such final things as death and judgment, heaven and hell, and the end of the world.

Blanke, O., "Out of body experiences and their neural basis," *British Medical Journal* 329 (2004): 1414–1415.

Ehrsson, H., "The experimental induction of out-of-body-experiences," *Science* 317 (2007): 1048.

The following articles represent some others that have also described a so-called out-of-body experience. However, the actual testimonies seem fundamentally different from what people describe when experiencing a sense of being separated.

Yu, K., C. Liu, T. Yu, et al., "Out-of-body experience in the anterior insular cortex during the intracranial electrodes stimulation in an epileptic child," *Journal of Clinical Neuroscience* 54 (2018): 122–125.

Kantaro, H., N. Shinoura, Y. Ryoji, and A. Midorikawa, "Dissociation of the subjective and objective bodies: Out-of-body experiences following the development of a posterior cingulate lesion," *Journal of Neuropsychology* 14 (2020): 183–192.

Lopez, C., and M. Elziere, "Out-of-body experience in vestibular disorders—A prospective study of 210 patients with dizziness," *Cortex* 104 (2018): 193–206.

Bos, E. M., J. K. H. Spoor, M. Smits, et al., "Out-of-body experience during awake Craniotomy," *World Neurosurgery* 92 (2016): 586.e9–586.e13.

Devinsky, O., E. Feldmann, K. Burrowes, and E. Bromfield, "Autoscopic phenomena with seizures," *Archives of Neurology* 46 (1989): 1080–1088.

Fang, T., R. Yan, and F. Fang, "Spontaneous out-of-body experience in a child with refractory right temporoparietal epilepsy," *Journal of Neurosurgery: Pediatrics* 14 (2014): 396–399.

Greyson, B., N. B. Fountain, L. L. Derr, and D. K. Broshek, "Out-of-body experiences associated with seizures," *Frontiers in Human Neuroscience* 8 (2014): 65.

Bateman, L., C. Jones, and J. Jomeen, "A narrative synthesis of women's out-of-body experiences during childbirth," *Journal of Midwifery Women's Health* 62 (2017): 442–451.

Chapter 16: Truth Is in the Eye of the Beholder

Timmerman, C., L. Roseman, L. Williams, et al., "DMT models the near-death experience," *Frontiers in Psychology* 9 (2018): 1424.

Greyson, B., "The Near-Death Experience Scale: Construction, reliability, and validity," *Journal of Nervous and Mental Disease* 171 (1983): 369–375.

Kondziella, D., J. Dreier, M. H. Olsen, "Prevalence of near-death experiences in people with and without REM sleep intrusion," *PeerJ* 7 (2019): e7585.

The following are published studies that included actual testimonies from people who have taken psychedelic substances. We found the experiences quite different from death-related experiences. For a summary of these findings, please see below and also refer to the supplementary sections of Parnia, S., S. G. Post, M. T. Lee, et al., "Guidelines and standards for the study of death and recalled experiences of death—a multidisciplinary consensus statement and proposed future directions," *Annals of the New York Academy of Sciences* 1511 (2022): 5–21.

Newcombe, R., "Ketamine case study: The phenomenology of a ketamine experience," *Addiction Research & Theory* 16 (2008): 209–215.

Simon, E., "Ketamine: Safe until it's not—a terrifying trip to the k-hole," *Journal of Emergency Medicine* 57 (2019): 587–588.

Pomarol-Clotet, E., G. D. Honey, G. K. Murray, et al., "Psychological effects of ketamine in healthy volunteers. Phenomenological study," *British Journal of Psychiatry* 189 (2006): 173–179.

Strassman, R. J., C. R. Qualls, E. H. Uhlenhuth, and R. Kellner, "Dose-response study of N,N-dimethyltryptamine in humans. II. Subjective effects and preliminary results of a new rating scale," *Archives of General Psychiatry* 51 (1994): 98–108.

Rock A., and C. Cott, "Phenomenology of N,N-dimethyltryptamine use: A thematic analysis," *Journal of Scientific Exploration* 22 (2008): 359–370.

Jansen K., *Ketamine: Dreams and Realities.* Sarasota, FL: Multidisciplinary Association for Psychedelic Studies, 2004.

Davis, A. K., J. M. Clifton, E. G. Weaver, E. S. Hurwitz, M. W. Johnson, and R. R. Griffiths, "Survey of entity encounter experiences occasioned by inhaled N,N-dimethyltryptamine: Phenomenology, interpretation, and enduring effects," *Journal of Psychopharmacology* 34 (2020): 1008–1020.

Newcomer, J. W., N. B. Farber, V. Jevtovic-Todorovic, et al., "Ketamine-induced NMDA receptor hypofunction as a model of memory impairment and psychosis," *Neuropsychopharmacology* 20 (1999): 106–118.

Gouzoulis-Mayfrank, E., K. Heekeren, A. Neukirch, et al., "Psychological effects of (S)-ketamine and N,N-dimethyltryptamine (DMT): A double-blind, cross-over study in healthy volunteers," *Pharmacopsychiatry* 38 (2005): 301–311.

The following publications highlight some of the other experiences (i.e., those that are not a recalled experience of death) that people can have over a period of days to weeks while in the intensive care unit. What has been labeled at times as a so-called hellish or frightening near-death experience (NDE) by some people in the past likely reflects these different experiences.

Doig, L., and K. Solverson, "Wanting to forget: Intrusive and delusional memories from critical illness," *Case Reports in Critical Care* 2020 (2020): 7324185.

Roberts, B., and W. Chaboyer, "Patients' dreams and unreal experiences following intensive care unit admission," *Nursing in Critical Care* 9 (2004): 173–180.

Roberts, B. L., C. M. Rickard, D. Rajbhandari, and P. Reynolds, "Patients' dreams in ICU: Recall at two years post discharge and comparison to delirium status during ICU admission. A multicentre cohort study," *Intensive Critical Care Nursing* 22 (2006): 264–273.

Papathanassoglou, E. D., and E. I. Patiraki, "Transformations of self: A phenomenological investigation into the lived experience of survivors of critical illness," *Nursing in Critical Care* 8 (2003): 13–21.

Duppils, G. S., and K. Wikblad, "Patients' experiences of being delirious," *Journal of Clinical Nursing* 16 (2007): 810–818.

Van Rompaey, B., A. Van Hoof, P. van Bogaert, O. Timmermans, and T. Dilles, "The patient's perception of a delirium: A qualitative research in a Belgian intensive care unit," *Intensive Critical Care Nursing* 32 (2016): 66–74.

Magarey, J. M., and H. H. McCutcheon, "'Fishing with the dead'—recall of memories from the ICU," *Intensive Critical Care Nursing* 21, no. 6 (2005): 344–354.

Löf, L., L. Berggren, and G. Ahlström, "Severely ill ICU patients recall of factual events and unreal experiences of hospital admission and ICU stay—3 and 12 months after discharge," *Intensive Critical Care Nursing* 22, no. 3 (2006): 154–166.

Further notes on hallucinogenic drug-induced experiences:

Thankfully, with the resurgence of interest in hallucinogenic drugs, more and more people have provided testimonies about their hallucinogenic experiences—and scientists and journalists alike have started to document these in the scientific and lay literature.

In 2017, Emilie Bowen, then a reporter working for *The Tab*—a British online newsmagazine that covers youth and student culture—published the testimonies of a group of people in their twenties in England who had undergone a DMT hallucination. Rachel, a twenty-two-year-old from Kent, told her, "I felt I was half in and half out of the room—I could sense I was there, but I was also in a place 'underneath' reality. My field of vision consisted of swirling colours as if it was a tie dye—within the swirls of colour I could see beings or entities that were coming back and forth from me. Whatever they were doing to me they were enjoying, I even felt they laughed and smiled at me at times, despite them having no mouth or way of making noise. For some reason, I felt that I had wet myself as I was coming out of the trip."

Alex, a twenty-one-year-old student, described seeing "a large purple lady who was dancing seductively." He explained that "as she [the seductive woman] faded away, I noticed other people were around me in what looked like an indoor garden, with fake grass on the floor and purple and yellow chairs and some tables dotted around." He added, "There were about four other people there, two were mingling and the other two looked like they were talking to me. I had to keep my eyes closed the entire time because if I opened them, I would see my mates' yellow faces floating in the middle of this indoor garden."

Tony, a twenty-year-old from Cardiff, recounted, "Everything I could see was extremely vivid and it felt like my body was vibrating, but in a comforting way. I took another hit, started to lean back and watched blades of grass appear to open out like concertina fans. As I hit the floor, I closed my eyes and felt my body melt away into the ground, but my mind was launched into hyper drive . . . The most incredible closed eye visuals, including shapes, colours and figures. Patterns and images were moving and intertwining. [There was] a long colourful winding road leading off into the distance, and a figure dancing seductively and drawing me towards her." He added, "I followed down the path for as long as I could, which felt like a lifetime, and saw flashes of places, objects and people from throughout my life along the way."

Interestingly, like Alex, Tony described seeing a "seductive woman," but also "closed eye visuals" and various figures. Although he used the word "figure," he was not referring to what people describe in death—a luminous, compassionate, loving being. He also referred to "flashes of places, objects and people" from his life. However, he had not experienced a reappraisal of his interactions with others throughout life. *In death, people don't just see flashes of objects or places and people; they reevaluate their entire life from the perspective of morality and ethics.* In turn, Rachel, one of the other people who had been interviewed, had said she felt "half out of the room," but she wasn't describing the experience of separation that people have in relation to death.

These cases illustrate the challenge with language and the limitations of making inferences simply based on a few words taken out of context. People may experience seeing places, objects, and people or being out of a room in multiple circumstances, including in a dream. It doesn't mean that every time we think back or see people from our past, it is a dream. *By relying on words alone without context, it is exceedingly difficult, if not impossible, to claim any experience is the same as or different from another.*

Equally, when the overall context of Tony's experience is considered, even the "flashes of places, objects and people" are different from the experience of dying. Yet someone could mislabel them as a "life review" and claim DMT produces the experience of dying. The same applies to Rachel's statement about feeling "half outside of the room." Someone could say this is the same as the sense of separation in relation to death. *This is how, without any stringent criteria or standards or objective scientific methods to analyze people's testimonies, people will inevitably label them as they wish based on their own beliefs.*

While these testimonies did not sound at all like the recalled experience of death, nonetheless our team of researchers at New York University decided to go one step further. We examined every single study—nine had been published between 1994 and 2019—in the scientific literature with testimonies of DMT or ketamine hallucinations. We included ketamine because some people have claimed this drug also mimics the experience of death. We wanted to better understand what those experiences were really like.

Based on all the buzz in the media and the claims made by some scientists like Dr. Timmerman, we expected to find similarities between DMT or ketamine hallucinations and the recalled experience of death. However, we did not. Instead of the universally consistent narrative arc of a separation from the body, evaluation and reappraisal of conduct in life, returning to a place that feels like home, and a decision to return to the body, hallucinogenic experiences typically involved haphazard geometric shapes, spirals, and unrelated beings, such as humanoids, elves, fish, seductive women, and aliens.

Overall, we found the reported DMT themes could be categorized as: *a) physical bodily changes* (like twitching or losing control), *b) becoming egotistical* (for example, feeling they had become godlike), *c) seeing random images (often geometrical)* or *d) beings* (for example, elves). The latter were different from the universally reported luminous, compassionate, loving being with enormous magnitude who guides people through an evaluation of their lives with humility. There were also *e) many other miscellaneous, varied, and unrelated themes.*

To illustrate the DMT experience better, we have provided a summary of all the major themes identified from the scientific literature—each illustrated with one or two quotes from people's published testimonies.

The first category involved physical bodily changes, including losing control of movements. One person said, "There's that feeling of the dip in the road, swinging on a swing, your stomach sinks. The flushes run through you, your legs twitch." One other person explained, "I looked up and saw how mechanical and essentially soul-less you were. Your movements were not your own, they were no longer smooth and coordinated."

A second category reflected egotistical perceptions. One person said, "I was essentially told that I was God."

The third category involved seeing random images or beings, such as elves. One person explained, "One of the elves made it impossible for me to move. There was no issue of control; they were totally in control. They wanted me to look here and there. That was all I could do." Another person said, "The 'elves' were prankish and ornery in their nature. There were four of them by the highway, they totally commanded the scene—it was their territory. They were about my height and held up placards."

Some saw other beings, too. One person said, "[Other intelligent figures] were aware of me, but not particularly concerned. It was like what a parent would feel looking into a playpen at his 1-year-old laying there." Another said, "[The being] was teaching me the rules/regulations of the NFL [National Football League]," while another said, "I thought about a missing zippo [cigarette] lighter for some reason and [the being] flashed to me where it was and after I came back, I went to that spot deep inside a couch and grabbed it perfectly, it was unreal."

The fourth category involved changes in emotions and sensations. One person said, "I had an overwhelming feeling of bliss and giggles while on the come up, and during the peak." By contrast, some described an inability to respond emotionally. One person explained, "I tried to get myself worked up over what I was seeing, but I just wasn't able to respond emotionally."

There were many other miscellaneous themes, too. One was *information overload.* One person said, "There is so much information coming at you at once that it is very hard to comprehend and make sense of at the time." Another involved a *change in focus*, as demonstrated by one person who said, "Things are coming into focus. I'm feeling human again. I had no idea what was going on." And another explained, "My mind was definitely at a different place, but it was commenting on the state as it was going on along."

One other theme involved *being in space or in a galaxy*, while another involved entering a *dimension of the universe through an object, such as a plant.* One individual said, "I was in the middle of the galaxy and there was no one to help me," while another explained, "I was greeted with a chrysanthemum [plant] in vivid lime-ish green and deep red, which then opened up into another plane, or dimension of existence, or some type of parallel conscious living part of the universe." There was also a theme about vibrations. One person said, "I was worried that the vibration would blow my head up. The colors and vibrations were so intense, I thought I would pop."

The ketamine experience was similar to that of DMT, but being a dissociative drug, there were more bodily distortions and self-seeing—autoscopic—experiences reported. The main themes identified were, *a) physical bodily changes, b) mental bodily misperceptions and distortions, c) becoming egotistical, c) seeing random images, d) beings*, or *e) a duplicate virtual self or selves.*

As regards *the first category—physical bodily changes*—people described losing control of their body and bodily movements. One person said, "[I am] not in control of my body, can't move." Another said, "I don't feel in control of my muscles anymore—like a zombie is a very good description of it. There is something making me just stay here. Something in my head is telling me I can't move." Whereas another said, "The will is there but difficult to get my legs to do what I want them to do." One other person said, "Everything takes a long time, for example moving my foot." Other similar examples were: "[I] feel like it would be impossible to stand up, [my] body feels like a ten-ton weight" or "my limbs feel like they've got a magnet and they're stuck to the arm of the chair like lead weights." People also described loss of coherent speech. One explained, "It was as though you were speaking gibberish," while another said there was "a delay between the thought and your mouth [moving]."

The second major category involved misperceptions and deformations of body image, such as seeing *elongated body parts.* One person said, "My body image was distorted beyond recognition—with fantastically elongated pipe cleaner legs and arms, spindly E.T. [referring to the fictional extraterrestrial character from the 1982 Hollywood movie of the same name]–like fingers, and morphing alien-insect head in the mirror . . . " One other reported, "My hands look small, but the fingers are really long."

Some bodily distortions—*as Dr. Blanke's patient had also described*—involved *enlarged or shrunken body parts*. One person said, "My hand looks like a midget hand . . . like a funhouse mirror effect." Another explained, "I feel like I'm shrunken inside," while someone else said, "My legs look very big and funny shaped, like another person's."

The third major category concerned egotistical perceptions. As with DMT, there was a very strong sense of becoming more egotistical, such as loving themselves, feeling godlike, or actually being God or some sort of goddess. One person said, "I felt God-like. I would love myself; it was great." Another said, "I was actually God. I distinctly felt the universe watching for my signal to see if it should cycle through itself once again, as it had an infinite number of times, or should it simply conclude." Still another said, "I was Isis herself, the virgin mother-goddess brooding lovingly over this world that I had created and was enfolding with arms like wings."

The fourth category concerned seeing random images or beings. As with DMT, people described various geometric shapes, neon lights, lasers, a kaleidoscope, a television screen, a fish scale, sandpaper, and so on. One person said, "Everything looks rounded," while another explained, "This is real light. I'm not talking about the God-light either, the one at the end of the tunnel and all that white light of the void stuff . . . This light in the body doesn't have that sense of meaning. It's more as if neon signs and lasers are actually inside you." *This person's testimony showed how the perception of light is very different between the hallucinogenic drug experiences and the recalled experience of death—even though both may use the word "light."*

Another person explained, "It's as though green emerald is like in and around me." Someone else said, "My mind bounces from a 1980s television screen, black and white with snow, to a kaleidoscope of colors." One person explained, "My visual world was made up of glistening fish scales." Another said, "You appear like a 2D image," and someone else explained, "I am feeling like I am made of sandpaper."

As with DMT, there were recollections of meeting a variety of entities. One person said, "I became aware of a 'ketamine creature' (Kreature) who was simultaneously some kind of spaceship, and it told me that the person I usually was in everyday life was also something like a four-dimensional 'badge' that was worn by some larger multi-dimensional entity."

A fifth category involved seeing duplicated—autoscopic—virtual images of the self. One person explained, "I actually saw myself split up into 3 different people, not in my mind either. I mean that I actually saw one of me to the right, and one of me on the left." Another explained, "I was sitting on the floor—but I was also sitting on the ceiling and on the walls, looking down and up and sideways at everything. Multiple perspectives—then suddenly more and more—I was sitting, standing, walking, flying, falling and totally still." Another explained, "K [ketamine] can split you into several personalities—different selves in one room."

One of the key distinguishing features between drug-induced experiences and the recalled experience of death that we discovered was their effect on people. *Although with drug-induced hallucinations people often experienced a sense of grandiosity—for example, thinking they are God or loving themselves—during the recalled experience of death, as one person explained, there is no ego left. This leaves people to experience a profound sense of humility.* They don't experience loving themselves or thinking they are a god or a goddess—even if after recovering from their life-threatening illness they may reexperience what it is like to have a sense of self with ego again. Even the highly luminous, compassionate entity with great magnitude, power, knowledge, and wisdom is always characterized with complete humility and without any ego.

Intriguingly, we found this sense of complete humility without any egotistic traits in spite of the fact that during day-to-day experiences none of us truly know what it is like to have a sense of self without any ego. Yet in death, together with experiencing a vast expansion of lucid consciousness and awareness, as well as a complete reappraisal of life from the perspective of morality and ethics, people also experience complete humility—their sense of self but without any ego. This became clearer to us as more and more testimonies were studied from across the world.

SECTION FIVE: WHAT IT ALL MEANS FOR US

Chapter 17: Putting It Together

Trujillo, C. A., R. Gao, P. D. Negraes, J. Gu, et al., "Complex oscillatory waves emerging from cortical organoids model early human brain network development," *Cell Stem Cell* 3, no. 25 (2019): 558–569.e7. doi: 10.1016/j.stem.2019.08.002. Epub 2019 Aug 29. PMID: 31474560.

Wilson, M. N., M. Thunemann, X. Liu, et al., "Multimodal monitoring of human cortical organoids implanted in mice reveal functional connection with visual cortex," *Nature Communications* 3, no. 1 (2022): 7945. doi: 10.1038/s41467-022-35536-3. PMID: 36572698; PMCID: PMC9792589.

For a brief discussion on cognitive dissonance, see *Encyclopedia Britannica*, [edition] ed., s.v. "cognitive dissonance."

Harmon-Jones, E., and J. Mills, "An introduction to cognitive dissonance theory and an overview of current perspectives on the theory," in E. Harmon-Jones (ed.), *Cognitive Dissonance: Reexamining a Pivotal Theory in Psychology*. Washington, DC: American Psychological Association, 2019, pp. 3–24.

Festinger, L., *A Theory of Cognitive Dissonance*. Evanston, IL: Row, Peterson, 1957.

Farahany, N. A., H. T. Greely, S. Hyman, et al., "The ethics of experimenting with human brain tissue," *Nature* 556 (2018): 429–432.

Farahany, N. A., H. T. Greely, and C. M. Giattino, "Part-revived pig brains raise a slew of ethical quandaries," *Nature* 568 (2019): 299–302.

Farahany, N. A., and H. T. Greely, "Advancing the ethical dialogue about monkey/human chimeric embryos," *Cell* 184 (2021): 1962–1963.